Design Reuse in Product Development Modeling, Analysis and Optimization

SERIES ON MANUFACTURING SYSTEMS AND TECHNOLOGY

Editors-in-Chief: Andrew Y. C. Nee *(National University of Singapore, Singapore)*
J.-H. Chun *(Massachusetts Institute of Technology, USA)*

Assistant Editor: S. K. Ong *(National University of Singapore, Singapore)*

Published

Vol. 1:   An Advanced Treatise on Fixture Design and Planning
Andrew Y. C. Nee, Z. J. Tao & A. Senthil Kumar

Vol. 2:   Integrated and Collaborative Product Development Environment: Technologies and Implementations
W. D. Li, S. K. Ong & A. Y. C. Nee

Vol. 3:   Fundamentals of Robotic Grasping and Fixturing
C. Xiong, H. Ding & Y. Xiong

Series on Manufacturing Systems and Technology — Vol. 4

Design Reuse in Product Development Modeling, Analysis and Optimization

S K Ong
National University of Singapore

Q L Xu
Nanyang Technological University, Singapore

Andrew Y C Nee
National University of Singapore

World Scientific

NEW JERSEY · LONDON · SINGAPORE · BEIJING · SHANGHAI · HONG KONG · TAIPEI · CHENNAI

Published by

World Scientific Publishing Co. Pte. Ltd.
5 Toh Tuck Link, Singapore 596224
USA office: 27 Warren Street, Suite 401-402, Hackensack, NJ 07601
UK office: 57 Shelton Street, Covent Garden, London WC2H 9HE

British Library Cataloguing-in-Publication Data
A catalogue record for this book is available from the British Library.

Series on Manufacturing Systems and Technology — Vol. 4
DESIGN REUSE IN PRODUCT DEVELOPMENT MODELING, ANALYSIS AND OPTIMIZATION

Copyright © 2008 by World Scientific Publishing Co. Pte. Ltd.

All rights reserved. This book, or parts thereof, may not be reproduced in any form or by any means, electronic or mechanical, including photocopying, recording or any information storage and retrieval system now known or to be invented, without written permission from the Publisher.

For photocopying of material in this volume, please pay a copying fee through the Copyright Clearance Center, Inc., 222 Rosewood Drive, Danvers, MA 01923, USA. In this case permission to photocopy is not required from the publisher.

ISBN-13 978-981-283-262-7
ISBN-10 981-283-262-9

Desk Editor: Tjan Kwang Wei

Printed in Singapore.

Preface

Today's market is characterized by intense competition in the global manufacturing environment. In order to succeed or even to survive, manufacturers must be able to deliver their products with speed, diversity, high quality, environment compliancy, and at low cost. Outstanding product design meeting the above-mentioned requirements will ultimately determine the final winner in this competition. In fact, there is a growing awareness of the vital role of product design for business success. Therefore, the desire for advanced design paradigms and powerful design tools has been persistent. On the design paradigm side, systematic product design and manufacturing principles are advancing, such as concurrent engineering (CE), Taguchi method, axiomatic design, theory of inventive problems solving (TRIZ), design for manufacturability and assembly (DFMA), mass customization, etc. On the practical design tool side, various computer-based systems and tools have been thriving, such as computer-aided design (CAD), computer-aided manufacturing (CAM), computer-aided engineering (CAE), product data management (PDM), expert systems, virtual reality (VR) and augmented reality (AR), etc. The last decade also witnessed the burgeoning of Internet technologies that facilitated a distributed, collaborative manufacturing environment.

Along with the flourishing methodologies and tools, comes information explosion such that the information base is too huge and volatile to be effectively managed. The dilemma faced by the designers in this information age is that a designer is "drown in data but thirsty for knowledge" [Rezayat, 2000]. There is an apparent appeal for more effective tools to manage the existing product information such that a designer can be supported by the relevant knowledge, at the right time, and in the right form.

In such a context, many researchers have been advocating design reuse as an effective knowledge management methodology. Design reuse is not aimed at delivering fancy solutions that will excite the customers and the designers themselves. Instead, it is a methodology to make good use of existing good ideas. The bottom line is: *we do not reinvent the wheel*. In the past few years, the authors of this book have been studying the applicability and effectiveness of design reuse in product development. These efforts involve different stages of product development. Nevertheless, the authors paid more attention to the early stages of product development because it is during these stages that the value of a product is largely determined. Moreover, with the increasing interest in product family design, significant effort has been made on applying the design reuse methodology in product family design.

The efforts made in these areas can be summarized in three areas, namely modeling, analysis, and optimization. Design reuse starts from existing product designs. Thus, modeling is necessary to capture the essential information in a form that is suitable for reuse. Modeling also involves the building of design system infrastructure to organize the information. Analysis refers to the activities to explore the design domain, identify useful knowledge pieces, and assemble them into reusable forms. Optimization mainly focuses on the design-by-reuse process, where solutions for new design tasks can be generated, evaluated, and optimized based on the design targets. An effective reuse system must integrate these interrelated processes to achieve the best efficacy.

This book is organized as follows. Chapter 1 provides an overview of the design reuse methodology and its relevancy to product development. The requirements to apply design reuse in product design are proposed. Chapter 2 presents a literature review of the legacy design reuse system and the computational tools to support the reasoning in design reuse. Chapter 3 is devoted to the information modeling issue. In particular, it presents the information modeling techniques to support design reuse. It introduces what information should be reused and how the information should be represented. The analysis issues are dealt with in Chapter 4. A central question in this respect is how to support the establishment of product platforms. A novel method based on neural networks is proposed. Chapters 5 ~ 7 present different aspects of the optimization issues. Chapter 5 presents a framework of optimization-based design synthesis approach for product configuration design. Chapter 6 summarizes the state-of-the-art techniques of cost estimation, where cost is an important criterion for optimization. Chapter 7 focuses on another optimization criterion, namely the product performance. A novel method based on axiomatic design is proposed to deal with this particular problem. Next, Chapter 8 presents an integrated design reuse system that amalgamates the technologies in design reuse and applies it to the design of product families. Finally, an online Web-based design reuse system is introduced in Chapter 9 to implement the embodiment and detailed design. In the book, various electro-mechanical products are used in the case studies to help the readers understand the methodologies.

This book covers topics in knowledge gathering, deployment, and utilization. It is intended to be useful for undergraduate and graduate students, and researchers in mechanical/industrial engineering and computer science to improve their understanding of the principles of product development. It can also be used as a reference book for practicing engineers and engineering managers to expand their visions of systematic product development and project management. The reader should have a basic understanding of mechanical products and systems. Fundamental knowledge of artificial intelligence is also helpful to understand the content of this book.

S.K. Ong
Q.L. Xu
A.Y.C. Nee
2007.12

Contents

Preface   v

1   Introduction   1
   1.1   Design Reuse – What and Why   2
       1.1.1   Types of design reuse   2
       1.1.2   The importance of design reuse   3
   1.2   Product Conceptual Design   6
       1.2.1   Product family design   8
   1.3   Major Issues in Design Reuse   10
       1.3.1   Design reuse process   11
       1.3.2   Product information modeling   12
       1.3.3   Product information analysis   13
       1.3.4   Design synthesis   13
       1.3.5   Solution evaluation   14
   1.4   Engineering Design Reuse Applications   15
       1.4.1   Design reuse in software engineering   15
       1.4.2   Design reuse in mechanical and electro-mechanical engineering   18
       1.4.3   Design reuse in manufacturing   20
   1.5   Barriers to Design Reuse   22
   1.6   Summary   24

2   Design Reuse Systems and Enabling Tools   27
   2.1   Engineering Design Reuse Approaches   27
       2.1.1   Case-based reasoning   28
       2.1.2   Catalog-based design   29
       2.1.3   Modular design   31
       2.1.4   Adaptable design   33
       2.1.5   Expert systems   35
       2.1.6   Innovative design using TRIZ   37
   2.2   Reasoning in Design Reuse   38
       2.2.1   Machine learning   38

		2.2.2 Data mining	40
		2.2.3 Design structure matrix	41
		2.2.4 Artificial neural networks	43
		2.2.5 Genetic algorithms	46
		2.2.6 Agent-based method	47
	2.3	Summary	49
3	Product Information Modeling		51
	3.1	Data, Information and Knowledge	51
	3.2	Information Modeling – State-Of-The-Art Review	53
		3.2.1 Content of information model	53
		3.2.2 Modeling languages	58
		3.2.3 Taxonomies	61
		3.2.4 Database system and web-based environment	63
	3.3	Function-Based Product Information Model	66
		3.3.1 A multiple facet product information model	66
		3.3.2 Representation of function using key element vector	69
		3.3.3 Function taxonomies	71
		3.3.4 An illustrative example	74
	3.4	Summary	78
4	Design of Product Platform		81
	4.1	Role of Product Platform	81
	4.2	Product Platform and Product Family Design	83
		4.2.1 A top-down perspective	84
		4.2.2 A bottom-up perspective	85
	4.3	Computational Tools for Product Architecture Building	87
		4.3.1 QFD-based approach	87
		4.3.2 DSM-based approach	88
		4.3.3 Heuristic and quantitative approaches	90
	4.4	Product Architecture Building Using Self-Organizing Map	91
		4.4.1 Introduction of SOM	91
		4.4.2 Function clustering based on SOM	94
		4.4.3 A case study	99
		4.4.4 Evaluation of the SOM method	103
	4.5	Other Relevant Issues in Product Platform Design	106
		4.5.1 Extraction of KCs as performance criteria	107
		4.5.2 Formation of component catalog	109
		4.5.3 Establishment of mapping route using correlation matrices	109
	4.6	Summary	111

5	Optimization in Product Design	113
	5.1 Introduction	113
	5.1.1 Weighted sum method	116
	5.1.2 Goal programming	117
	5.1.3 Multi-level programming/rank ordering	118
	5.1.4 Genetic algorithms	118
	5.2 Automated Design Synthesis	121
	5.2.1 Configuration design	121
	5.2.2 Design synthesis techniques	122
	5.3 Multi-objective Struggle Genetic Algorithm Design Synthesis	128
	5.3.1 Problem formulation	128
	5.3.2 The MOSGA algorithm	131
	5.3.3 Implementation of MOSGA in product configuration design	133
	5.3.4 Precautions and limitations	139
	5.4 Post-optimal Solution Selection	140
	5.5 A Case Study	142
	5.5.1 Experience-based design	144
	5.5.2 Product design using the design reuse approach	146
	5.5.3 Comparison of the two methods	151
	5.6 Summary	151
6	Cost Estimation in Product Development	153
	6.1 Introduction	153
	6.2 Product Development Cost	155
	6.2.1 Cost structure	155
	6.2.2 Cost modeling techniques	158
	6.3 Cost Estimation in Product Family Development	166
	6.3.1 Commonality index	167
	6.4 An Empirical Cost Model for Design Reuse	169
	6.4.1 Fixed cost	170
	6.4.2 Development cost	171
	6.4.3 Component cost	171
	6.5 Summary	173
7	Product Performance Evaluation	175
	7.1 Introduction	175
	7.1.1 Relating performance to design parameters	175
	7.1.2 Aggregating performance criteria	177
	7.2 Robust Design	178
	7.3 The Information Content Assessment Method	182
	7.3.1 Background – information axiom and information content	183
	7.3.2 The information content assessment process	186

7.3.3 Establishing system range from existing products	187
7.3.4 Assessing information content	193
7.3.5 Precautions and limitations	197
7.3.6 A case study	198
7.4 Summary	203
8 A Product Family Design Reuse Methodology	**205**
8.1 Introduction	205
8.1.1 Scale-based approach	206
8.1.2 Model-based approach	211
8.1.3 Graph-based approach	211
8.1.4 Module-based approach	211
8.2 An Integrated Design Reuse Process Model	212
8.2.1 Product information modeling	213
8.2.2 Knowledge extraction	214
8.2.3 Design synthesis and evaluation	216
8.3 A Web-Based Product Family Design Reuse System	216
8.4 Design of Cellular Phone Product Family	222
8.4.1 Settings	222
8.4.2 Results	226
8.4.3 Analysis	228
8.5 Design of TV Receiver Circuits	229
8.5.1 Settings	229
8.5.2 Solution generation and results	232
8.5.3 Comparison	233
8.6 Summary	234
9 Design Reuse for Embodiment and Detailed Design	**237**
9.1 Introduction	237
9.2 Online Design Reuse System	239
9.2.1 System architecture	239
9.2.2 Product information representation	241
9.3 Embodiment Design	241
9.3.1 Product case retrieval method	242
9.3.2 Optimal search for alternative solution	247
9.3.3 Exhaustive search	254
9.3.4 GA-based search	255
9.3.5 Solution generation in washing machine design	260
9.4 Detailed Design	263
9.4.1 Architecture of the detailed design transformation	263
9.4.2 Feature-based parametric modeling	265
9.4.3 Product family and variant method	266

9.4.4 Operation of detailed design reuse	267
9.4.5 System implementation	271
9.5 Summary	271
Bibliography	273
Index	293

Chapter 1

Introduction

Product design is concerned with the design of products to achieve the desired functionality, where functionality may involve a number of features, such as usage, safety, durability, aesthetics, and social and environmental issues. To cater to the multitude of functionality issues is not a simple task. Moreover, the intense competition in the global manufacturing environment makes the task even tougher. In order to succeed or even to survive, a manufacturer must be able to deliver their products with speed, diversity, high quality, and at low cost. These present a formidable challenge to engineers in product development. There is an apparent appeal for powerful design assisting tools.

In the past few decades, many powerful computer-aided technologies have emerged, such as computer-aided design and manufacturing (CAD/CAM), computer-aided process planning (CAPP), computer-aided engineering (CAE), etc. Artificial intelligence (AI) technologies find broad applications in engineering design to facilitate quicker and smarter decisions. Product data management (PDM) systems are widely used to manage the vast pool of product information throughout the product life-cycle (PLC). Despite these efforts, designers still face tough situations when making decisions at the early design stage. The need for effective design tools to support new product development and ensure product continuity is paramount. This is where design reuse can play an important role. This chapter introduces the rationale of design reuse and the legacy systems in various domains. The major issues to be addressed in design reuse are discussed.

1.1 Design Reuse – What and Why

1.1.1 *Types of design reuse*

Design reuse involves various activities that utilize existing technologies to address new design problems. The ultimate aim of design reuse is to assist the designers to develop products that maximize the value of the designed artefacts with minimal resources, cost and effort [Sivaloganathan and Shahin, 1999]. Basically, reuse can be divided into three types with respect to the objects to be reused.

(1) End-of-life product reuse (Type I), which refers to the reuse and recycling of obsolete products or components such that the components or materials can return to the PLC. This results in savings of natural resources and reduction of environmental impacts [Hata *et al.*, 1997; Kimura *et al.*, 1998].

(2) Reuse of existing manufacturing resources (Type II). Manufacturing processes inevitably consume energy and resources, especially when the manufacturing equipments have to be redesigned, upgraded, or reconfigured. Therefore, the configurations of different products must be designed in such a way that the production processes can be reused and shared. Production cost can be reduced through the utilization of existing manufacturing resources to accommodate the changing production requirements [Kimura and Nielsen, 2005].

(3) Reuse of product information and design knowledge (Type III). This type of reuse is a prerequisite for the earlier two types of reuse because design ultimately determines the extent to which the products and the manufacturing resources can be reused. In other words, effective reuse of available resources could not be achieved unless the products are designed to be reusable.

These three types of reuse roughly correspond to the reuse activities in the different stages in the PLC, as shown in Figure 1.1. This book focuses on the third type of design reuse, *i.e.*, the various approaches that support the utilization of knowledge gained from previous design activities. This is based on the belief that knowledge/information reuse enables the reuse of components and manufacturing resources, and hence is essential to sustainable design and manufacturing.

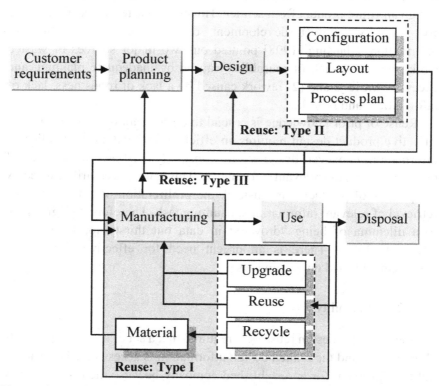

Figure 1.1 Type of design reuse in the product life-cycle

1.1.2 *The importance of design reuse*

Before an extensive study of the tools to assist design reuse, three fundamental questions have to be answered.
- Why is design reuse necessary?
- Is it possible to apply design reuse?
- Is design reuse methodology effective in product design?

1.1.2.1 Necessity

In today's market, no enterprise can afford the time and resources to design an entire product from scratch. However, wasted work still plagues contemporary product development due to the unavailability of information. Clausing [1998] pointed out two major sources of wasted work, *viz.*, (1) a lack of reusability caused by inadequate planning and excessive variety, and (2) rework caused by a lack of robustness, lack of information, and mistakes.

Reuse of prior knowledge is crucial to design rapidity and continuity. Effective product design requires an efficient retrieval and utilization of information. However, designers are constantly frustrated by the lack of means to access the relevant information. This is not necessarily caused by the paucity of product data. Instead, the proliferation of data makes the retrieval of relevant information a daunting task. Therefore, designers are in a dilemma of being "drowned in data but thirsty for knowledge" [Rezayat, 2000]. There is an urgent need for effective information management based on design reuse.

1.1.2.2 Applicability

In order to apply design reuse, it is required that a set of designed products already exist and the related design information is accessible. This should not be a problem for an established company because there is usually a pool of designed products. Typical in the industry, product development is evolutionary rather than revolutionary. According to statistics, only about 20% of an OEM's investment is on new design while about 80% is on the reuse of existing products, with or without modification [Rezayat, 2000]. Thus, design reuse can be applied in a broad variety of industries. Figure 1.2 shows a typical product development road-map. The horizontal axis is the time divided into years and quarters. A family of products (denoted by the hexagons) is distributed in three tiers, namely the high tier, mid tier and mass tier, according to the market segmentations shown on the vertical axis. The curve on the right shows the production volume in the different market segmentations. From this road-map, it can be observed that there is a constant migration of technologies from the higher end to the lower end as time proceeds. This ensures the continuation of product development within a corporation.

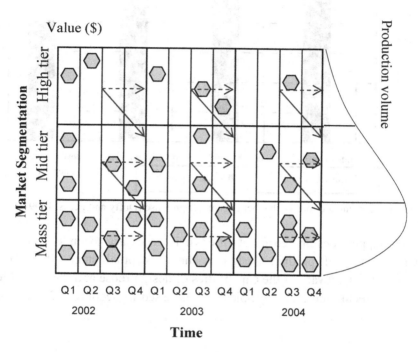

Figure 1.2 A typical product development road-map

1.1.2.3 Effectiveness

The effectiveness of design reuse should be validated by the improvements in the key factors of production, namely cost, quality, and time-to-market. It is expected that production efficiency can be increased because the designers do not have to start from scratch. Product quality can be improved by reusing the sub-systems or components which quality and validity have been proven. In addition, the outcome of the design can be better predicted, which is valuable to the early decision-making stage. By properly reusing existing technologies, significant benefits can be achieved with respect to cost, time, product quality, and performance [Duffy and Ferns, 1999] (Figure 1.3).

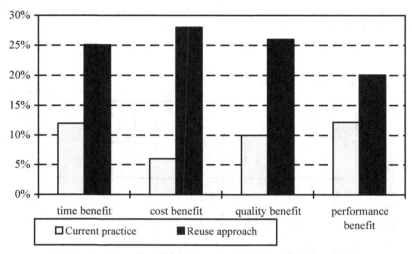

Figure 1.3 Current and foreseeable benefits of design reuse [Duffy and Ferns, 1999]

To support design reuse activities, it is necessary to understand the characteristics of product design at the conceptual stage. It is also important to be aware of the capabilities of design reuse and the available tools and techniques. These topics are discussed in Sections 1.2 and 1.3, respectively.

1.2 Product Conceptual Design

Among the several stages of product design, which usually encompass requirement analysis, conceptual design, embodiment design, and detailed design, the conceptual design stage is of paramount importance. This can be shown with two observations. Firstly, the conceptual stage allows for high design freedom, *i.e.,* the designer is less constrained to make decisions at this stage. Secondly, the cost of a product is largely determined at this stage. It is estimated that about 75% of the manufacturing cost is committed by the end of the conceptual stage [Ullman, 1997]. In the subsequent stages, it becomes increasingly difficult and costly to compensate for the initial flawed designs. This situation is illustrated in Figure 1.4.

Conceptual design is a design process that involves intense decision-making. A systematic, procedural process model must be developed to manage these decision-making activities. A few notable design theories that have dealt with this problem include the systematic approach [Pahl and Beitz, 1996], total design [Pugh, 1991], robust design [Clausing, 1994], the theory of inventive problem solving (TRIZ) [Altshuller, 1984], axiomatic design [Suh, 2001], *etc.*

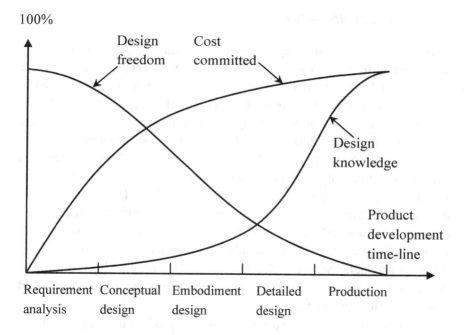

Figure 1.4 A product life-cycle viewpoint of design freedom, product cost and knowledge availability

At the early design stages, decisions have to be made on the project definition, design specifications, concept generation, concept evaluation, and the preliminary production issues. The effectiveness in carrying out these activities depends a lot on the availability of information, and the way in which the information is processed. However, the conceptual design stage is characterized by information deficiency and uncertainty. It is not until the final design stages that design knowledge becomes substantial. A paramount problem is how to carry out design based on the limited amount of information at the conceptual stage. Collection of information from existing products is a possible way to solve the problem. However, product information is highly unstructured and appears in diverse forms. Significant effort is required to capture product information, and utilize the information in new design problems.

1.2.1 *Product family design*

In conceptual design, the target can be the design of a single product or a set of related products, *i.e.*, a product family. A product family refers to a group of related products that share common technologies and address a series of market segmentations [Meyer and Lehnerd, 1997]. Product family design is a nascent but rapidly maturing field of research [Simpson, 2004]. The rationale of product family design is to provide product variety while maintaining production efficiency [Pine, 1993a]. Product variety is defined in terms of customer requirements, which are addressed by variegated product performance. Thus, a product family has to be designed to cover a ranged set of performance requirements. At the same time, production efficiency has to be ensured through considering commonality, compatibility, standardization and modularity among different products [Meyer and Lehnerd, 1997]. This is achieved through developing common technologies and components, which can be shared among different products.

Product family design and platform-based product development have been widely applied in the industry. A few successful projects in the industry are presented next.

Black & Decker – Power tool family Black & Decker used modularization to produce the entire range of the power tool products using standardized components, such as motors, bearings, switches, cord sets, cartons, fasteners, *etc.* [Lehnerd, 1987]. The product variety increased while production cost and lead-time decreased drastically.

Sony – Walkman family Sony introduced the Walkman with a variety-intensive strategy [Uzumeri and Sanderson, 1995]. The company built its models on common platforms and used modular design to provide variety. Sony also adopted an incremental innovation strategy and targeted different market niches with different models.

Swatch – Swiss watch family Swatch produced a large variety of watches by using a modular design and combining different standard chunks. A number of watch models with different configurations (hands, faces, wristbands, etc.) were created with a relatively small selection of movements and cases [Ulrich and Eppinger, 2004].

Xerox – Copier & typewriter family Xerox employed a modular design strategy in the design of its typewriters. Xerox also focused on reuse in product development to maximize the number of modules that are carried over from one generation to the next. This helps to lower development costs and decrease the lead-time [Erixon, 1996].

Boeing – Commercial airplane family Boeing developed its commercial airplanes based on a strategic 'stretching' of the basic models to accommodate different load capacities (*e.g.*, passenger number and cargo weight) and flight range [Sabbagh, 1996].

The major concern in product family design is the management of the trade-offs between product commonality and product performance. Usually, an increased commonality leads to a higher production efficiency, but at the expense of individual product performance. Decisions have to be made at the early design stages about (1) the proper divisions of the market segmentations, (2) the structure and content of a product platform, (3) the attributes of the common components under the product platform, and (4) the optimal combination and adaptation of components. Thereafter, it is important to evaluate (5) the effectiveness of the product family with respect to cost and product performance.

Information deficiency and uncertainty is a big hindrance to product family design. Usually, a designer is faced with immense freedom to develop a product family. It is not trivial to set the right parameters as a good starting point, *e.g.*, little is known about the consequences of setting a parameter at a specific value. Therefore, it is necessary to find ways to collect the relevant information and use it to ensure design optimality. The design reuse methodology presented in this book is one of such efforts to meet this challenge.

1.3 Major Issues in Design Reuse

Design reuse encompasses the design activities covering the entire product design processes. It involves various issues ranging from theoretical to practical, and technical to managerial. Sivaloganathan and Shahin [1999] summarized the developments in design reuse and classified the efforts into seven categories.

(1) Focused innovation for product development, which is largely a management issue addressing the planning of products and market segments to increase the market share.
(2) Cognitive studies on reuse, which emphasize the identification of patterns in the artifact information and use them to answer specific questions.
(3) Computational perspective of design reuse, which focuses on the reasoning of information retrieval, utilization, and adaptation. AI technologies play an important role in this area.
(4) The use of standard components, which emphasizes the cost effectiveness in manufacturing and efficiency in assembly.
(5) Design reuse tools, which refer to various stand-alone assisting tools that enable more efficient information manipulation, such as the market segment table, design structure matrix (DSM), *etc.*
(6) Design reuse systems, which integrate multiple tools to carry out the design reuse processes of interest. The systems may differ in scope and complexity based on the nature of the target problem.
(7) Problems of overuse, which identify the limitations of design reuse and the pitfalls for applying design reuse.

Other interesting issues in design reuse include environmental concerns, organizational strategy, and educational issues. Although these issues are comprehensive to a certain extent, they are presented in different levels of abstraction and lack a consistent framework. With an emphasis on the efficacy of applying design reuse to product design, this book concentrates on the issues related to the design reuse processes, which include a design reuse process model and the techniques associated with each and every individual process. These are discussed in the following sub-sections.

1.3.1 *Design reuse process*

To properly organize the design reuse process, a comprehensive design reuse process model is required. Systematic design reuse method involves two interrelated processes: information collection and information reuse. The former refers to design-for-reuse, which involves information modeling and information processing to identify relevant knowledge. The latter refers to design-by-reuse, which aims at the effective utilization of the information. Design-by-reuse is mainly concerned with information retrieval, solution synthesis and evaluation. Finger [1998] has identified four issues concerning the design reuse process, *viz.*, representing, capturing, organizing, and retrieving. This division is similar to the processes presented in traditional case-based reasoning (CBR), which is centered on '4Rs', *viz.,,* retrieve, reuse, repair, and retain [Watson, 1999].

These process models have been criticized for being based on a non-holistic model, *i.e.*, the overall design process has not been well-organized [Smith, 2002]. A relatively complete design reuse process model was proposed by Duffy *et al.* [1995]. It consists of three processes and six knowledge resources (Figure 1.5). This model is enlightening in that it identifies the role of knowledge resources, the flow of information, and the requirements of information processing. An effective design reuse system has to provide tools to facilitate the design processes and manage the relationships between the knowledge resources.

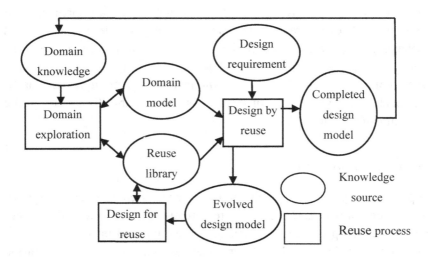

Figure 1.5 A design reuse process model [Duffy et al., 1995]

1.3.2 Product information modeling

The representation of the product information directly influences the effectiveness of design reuse. Since the product data are inherently heterogeneous and volatile in nature, the representation scheme has to deal with information completeness, conciseness, and integrity. The exchangeability of product information is also an important issue to be considered for collaborative design. Generic modeling languages, such as UML (Unified Modeling Language), CML (Compositional Modeling Language), STEP (Standard for the Exchange of Product model data), XML (eXtensible Markup Language), *etc.*, may facilitate the process. These modeling languages provide a common syntax with well-defined semantics to model a broad variety of physical processes and objects. However, their applications have been restricted by the efficacy to deal with representation flexibility and rigor.

One important aspect of information is product function. The use of function effectively separates the design intent with the physical implementation, and hence, design is partially exempted from early engagement to specific physical structures. Function-based product design has been recognized as an effective means to conceptual design. Therefore, the representation and subsequent reasoning about function has been under extensive study [Umeda *et al.*, 1990; Iwasaki and Chandrasekaran, 1992; Gorti and Sriram, 1996; Qian and Gero, 1996; Pahl and Beitz 1996; Roy *et al.*, 2001]. Relevant research issues include the representation scheme based on functions and flows, the building of function structures, the usage of taxonomy, the relationships between function, form and behavior, *etc*. Production information representation is further discussed in Chapter 3.

1.3.3 *Product information analysis*

Product information that is collected based on the representation schemes discussed in the earlier section is not necessarily reusable. Information is reusable if it can be easily retrieved and assembled to support solution generation. A notable difficulty faced by product engineers is that the information at hand lacks association with other types of information, and hence could not be directly reused. Techniques are required to transform product data into reusable forms. Thus, information analysis is another important issue in design reuse.

Information analysis usually involves the assignment of rules and the recognition of design patterns from the original data. This may be carried out at different levels of the product development, such as customer requirement analysis [Jiao and Zhang, 2005], building of product architectures [McAdams *et al.*, 1999; Hölttä *et al.*, 2003], process planning [Treleven and Wacker, 1987], and logistics [Kim *et al.*, 2002; Huang *et al.*, 2005]. Various AI techniques have been applied in such an effort, such as machine learning, data mining, neural networks, and heuristic methods. Chapter 4 discusses this issue in greater detail.

1.3.4 *Design synthesis*

Design synthesis has different implications according to the nature of the target problem. For example, Chakrabarti [2002] divides synthesis into five levels, *viz.*, synthesis as designing, synthesis as problem solving, synthesis as design solution generation, synthesis as design problem and solution generation, and synthesis as exploration. Based on the design reuse framework presented in this book, design synthesis refers to the generation of solutions based on reusable components. In order to do so, three factors have to be considered, *viz.*, (1) knowledge of the components or artifacts, (2) methods and techniques to implement the solution generation activities, and (3) knowledge of how these methods and techniques can be carried out.

Methods to carry out synthesis differ in scope and level of automation. Typically, design synthesis is carried out manually, or through the interactions between humans and computers. These methods are applicable to problems where simple retrieval and adaptation are involved. On the other hand, to achieve a more efficient product design, automated compositional synthesis is required. Automated design synthesis is especially useful for solving large combinatorial problems, such as configuration design. To do so, a proper formulation of the design problem is necessary so that the computers can read design inputs, visit relevant data, and compute and present the synthesized results. Usually, automated design synthesis involves a set of predefined design objectives based on which the computers can evaluate the candidate solutions to search for the optimal ones. Design synthesis can be carried out using various computational tools, such as agent-based methods, genetic algorithms (GA), simulated annealing (SA), branch-and-bound method, *etc*. Chapter 5 presents the various methods to carry out design synthesis based on optimization.

1.3.5 *Solution evaluation*

The feasibility and optimality of a design concept is assessed using the concept evaluation schemes. The major difficulty in this process is that a mathematical model is often out of the question due to the complexity of the problem. In particular, two obstacles are prominent. Firstly, evaluation usually involves multiple criteria that are inherently incommensurable. The designer can aggregate the criteria into a multivariate utility function, or alternatively he/she can carry out the evaluation based on multi-objective optimization. However, a multivariate utility function is not easy to formulate; and the trade-offs are hardly manageable when many objectives are involved. Secondly, the logical management of the evaluation process is not trivial. The designer has to identify sufficient information and develop logical steps to compute the objective functions.

Due to these obstacles, early stage solution evaluation is difficult and has been relying on intuition and experience. This strategy is far from efficient and reliable due to the limitations of human capacity. Efforts have been made to tackle this problem by using systematic methods with computer support. For example, the quality function deployment (QFD) method has been widely applied to translate the customer needs into design specifications, and to further identify the most promising conceptual solutions. Another influential body of researches lies in quality engineering, and most notably the Taguchi method, which adopts prescriptive strategies in the design and manufacturing processes to ensure product quality. In Chapter 7, the issues in product quality evaluation and control are discussed.

Based on the discussions, this book is aimed at addressing these issues using an integrated framework with the supported of computational tools.

1.4 Engineering Design Reuse Applications

Design reuse is a multi-disciplinary research topic. This section discusses the applications of design reuse in different engineering domains, including software engineering, mechanical and electro-mechanical engineering, and manufacturing. Although this book focuses more on mechanical and electrical product design, the design reuse rationale is applicable to multiple disciplines. Thus, design reuse applications in different areas are complementary to each other, and can expand the visions of engineers from different fields.

1.4.1 *Design reuse in software engineering*

Design reuse as a research topic stems from computer science and software development. Reuse in software engineering has been successfully implemented at different levels: from the low-level code reuse to component reuse, and high-level system and project reuse. Reusable objects include code segments, components, structures (skeletons), documentation, report, test component, plans, *etc.*

Object-oriented (OO) design is a main enabler of software reuse. It is basically a software decomposition technique, which elicits systematic thinking to facilitate reuse, and provides environments and languages for programming. As compared with the classical functional design, OO design is based on a modular decomposition of a system and the concept of classes of objects, instead of the functions that the system performs [Jette and Smith, 1989]. An important implication of this difference is that OO design features a high-level abstraction at design time. This is a significant improvement to the classical procedural, flow-oriented design approach, in which reusability is not considered.

The most important concept in OO design is *Class*, which is an abstraction of a set of objects that share the common characteristics, structures, and operations. A class defines a data structure which consists of a set of attributes used to describe an object, and a set of methods used to define the operations that this object carries out. The key features of OO languages can be found in [Jette and Smith, 1989] and include:

Abstraction, which refers to the representation of objects with a high level of generalization. In other words, only information related to the nature of an object is retained, while the other non-essential information is ruled out. The class is a resulting data structure of abstraction. Class can be used to represent non-structured data as well as structured data.

Inheritance, which enables a class, called an heir/descendant, to obtain some of its features from another, called a parent/ancestor. Thus, the infrastructure of an OO system embraces a hierarchal structure of objects such that the descendants in the hierarchy can reuse the attributes and methods of the ancestors.

Polymorphism, which is an important consequence of inheritance. It allows descendants to override some of the attributes or methods of the ancestors provided that some prescriptive measures have been taken to define of the original data structure.

Encapsulation, which is a mechanism of information hiding. Encapsulation separates the external aspects of an object from the internal implementation details. The external aspects are accessible to other objects while the internal implementation are hidden form other objects. This ensures that only methods on the class could access its implementation [Meyer, 1994].

Software reuse has been extensively studied and widely adopted. A few successful design reuse applications are presented next.

Toshiba Fuchu Software Factory A standard life-cycle model is used to produce software at Fuchu. This factory embraces a motto of 'Promote Reuse', and reuse is fully supported by management. Reusable software were developed and stored in a large reuse library. Every project is encouraged, and in fact mandatory to review possible candidates for reuse at the start and throughout the project development. These practices have helped the company to achieve a 14% gain in productivity annually [Rada, 1995].

HP Corporate Engineering Software Reuse Program This program has been proposed to provide solutions to consulting, training, methods development, and pilot projects [Griss, 1991]. A hypertext-based software reuse tool was developed to integrate tools, and maintain links between all software work projects. A prototype called Kiosk provides a hypertext framework for manipulating libraries, such as InterViews, or other C and C++ libraries.

Reuse at IBM IBM established a reusable part center at Boblingen, Germany in 1987, which aimed at production of highly generic reusable software components for worldwide use within IBM [Wasmund, 1993]. Reusability is considered throughout the cycle of software development, which involves five major steps: (1) define the goal, (2) determine critical success factors, (3) define the required activities, (4) validate the plan, and

(5) execute the activities. These efforts have resulted in a tripling of reuse rate in IBM, Boblingen.

Reuse at NEC The NEC software engineering laboratory made a retrospective analysis of its business applications and recognized 32 logic templates and 130 common algorithms [Rada, 1995]. These templates and algorithms are classified and stored in a reuse library, which was in turn integrated into NEC's software development environment. A 7:1 improvement in productivity and 3:1 improvement in quality was reported.

Reuse in software engineering is both challenging and rewarding. To succeed with reuse, engineers and managers must see it from a systematic perspective. The design reuse practices must be formalized by including support for reuse in software development methods, tools, training, incentives, and measurements [McClure, 1997].

1.4.2 *Design reuse in mechanical and electro-mechanical engineering*

As compared to software design reuse, design reuse in mechanical and electro-mechanical design is more complicated. An apparent reason for this is that the products involving solid models of components are much more difficult to modify and customize. In general, the design of these components/products involves a more complicated process of mapping from the functional domain to the physical domain. Moreover, the functionality of the products is more complicated as a result of different combinations of components. Hence, the compatibility among interacting components requires more consideration. Specifically, the interface must be designed properly to avoid possible conflicts.

Modular design is an established method in this domain. The most successful modular product is probably the personal computer (PC). A PC usually consists of a set of functional components, such as central processing unit (CPU), storage devices, memory, power supply, graphic processor, input devices, *etc*. Standard interfaces are developed to connect these components so that they can be easily mounted on a main board during the assembly of a PC system.

The Cambridge Engineering Design Centre (EDC) in UK focused on the development, validation and dissemination of advanced design methods for mechanical systems [Clarkson, 1998]. The major research methods include: (1) functional synthesis, (2) embodiment generation, (3) design optimization, and (4) designer guidance. Reuse of knowledge is a central theme that unites these methods. Applications have been reported in aerospace, healthcare, and other special projects.

A-Design combines multi-objective optimization with a multi-agent approach for automated design synthesis [Campbell *et al.*, 1999, 2000]. It is capable of accepting changing design inputs and decision-making based on previous experience. These capabilities make the system intelligent and adaptive to dynamic environments. It has been used to design weighing machines and MEMS accelerometer.

Schemebuilder© [Cousell *et al.*, 1999] is a commercially available system for mechatronics conceptual design. The system is developed based on comprehensive knowledge representation and function-means tree to enable mapping between the functional domain and the physical domain. AI techniques have been adopted to support design reasoning. However, the interface design and the compatibility of different sub-systems have not been addressed in this system.

Other applications of design reuse in mechanical and electro-mechanical products include the design history tool [Chen, *et al.*, 1990], FAMING for design innovation [Faltings, 2002], issue-based information systems [Conklin and Burgess-Yakamovic, 1991], the multiple viewpoint modular method [Smith, 2002], LearninIT in circuit breaker design [Stahovich, 2000], *etc.* Reuse has been implemented at different levels in these systems. Usually, a system only provides partial support to reuse, *i.e.,* only a particular type of information is reused, such as layout, design history, modules, and reasoning. It is desirable to extend reuse as much as possible through more effective knowledge manipulation techniques.

1.4.3 *Design reuse in manufacturing*

Reuse in manufacturing can be applied at three levels. The methods involve both managerial and technical aspects, which are discussed next.

1.4.3.1 Management level reuse

With the increasing product variety and complexity, the number of process variations increases in terms of machine tools, fixtures, setups, cycle times, and labor [Wortmann, *et al.*, 1997]. Process variety may have an adverse impact on production efficiency and cost because it introduces significant constraints to production planning and control. Therefore, a paramount problem is how to effectively manage the process variety according to the product variety based on the existing operations and manufacturing resources.

The management of product variety has been supported by various tools and systems, such as PDM, bill-of-materials (BOM), Generic bill-of-materials (GBOM), *etc*. These tools and systems further facilitate the reduction of components in the inventory and resources to handle the components [Fisher *et al.*, 1999]. Moreover, the manufacturers can better control their resources by designing optimal supply chain configurations. Examples of resource management include the enterprise resource planning (ERP), material requirement planning (MRP), manufacturing resource planning (MRP II), *etc*.

However, the information provided by these systems may not be effectively reused. This is attributable to the limited capacity of the systems and the inefficiency of the tools to use information. Therefore, data mining technologies have been applied to extract useful data patterns for a more effective information reuse, such as graph-theoretic approach [Romanowski and Nagi, 2002], text mining and graph matching [Jiao *et al.*, 2005], and associating rule mining [Jiao *et al.*, 2005].

1.4.3.2 System level reuse

Manufacturing systems have different forms which are suitable for different production rationales. Accordingly, the reusability of the systems is different. Table 1.1 summarizes the features of different manufacturing systems in terms of their reusability.

Dedicated manufacturing lines (DML) A DML is a fixed production line that produces identical or similar products in high volumes. Typically, each dedicated line is designed to produce a single product or part. In DML, the production rate is very high. However, to produce components with different features, and at low volume, the reusability of DML is minimal.

Group technology (GT) GT is based on the rationale of grouping parts with similar production processes [Rolstadas, 1991]. This grouping results in cellular manufacturing structures, with each cell dedicated to a cluster of similar parts, and operates much the same as in the traditional DML. GT is capable of reducing setup time, inventory, and tool usage. It is flexible and allows organizations to be more responsive to the market changes.

Flexible manufacturing systems (FMS) FMS is a manufacturing philosophy and technology that is capable of producing product variants, with shifting volume and mix [Hopp and Spearman, 2001]. From the perspective of producing a variety of products, FMS is a reusable system. However, the throughput of FMS is low, especially when considering the high equipment cost.

Reconfigurable manufacturing systems (RMS) RMS has its significance in changing the system itself to adapt to the changing product variations. It combines the high throughput of DML with the flexibility of FMS. This has been made possible by: (1) a system composed of machines that are adjustable in scale and capacity to meet the changing requirements of the market, and (2) a manufacturing system that has the capacity to produce all members of the part family [Koren, *et al.*, 1999]. Thus, reuse is an inherent nature of RMS.

Table 1.1 Features of different manufacturing systems

Type	Flexibility	Scalability	Machine structure	Throughput	Tools	Reusability
DML	No	No	Fixed	High	Multiple	Low
GT	Yes	No	Fixed	High	Multiple	High within manufacturing cells
FMS	Yes	Yes	Fixed	Low	Single	Reusable for different products
RMS	Yes	Yes	Adjustable	High	Multiple	High

1.4.3.3 Reuse of machine tools

The capacity and adaptability of machine tools show different levels of reusability. Machine tools that are used in the DML are highly specific, and hence are not reusable for other production requirements. On the other hand, CNC machines are all-purpose machines that can accommodate a wide range of production requirements. To increase the reusability of machine tools, a modular structure has been widely adopted. A modular structure of the machine tools is defined as a set of pre-fabricated standard components that can be assembled rapidly into a variety of design configurations to address the production requirements. In fact, RMS could not have been effectively applied without the modular structure of machine tools. Successful applications of modular structures have been reported, such as modular jigs and fixtures [Nee, *et al.*, 1995, Rong and Zhu, 1999], and modular machine tools [Stake and Blackenfelt, 1998; Koren and Kota, 1999].

1.5 Barriers to Design Reuse

Although the design reuse methodology has many attractive features, such as the potential for cost savings, quality enhancement, time reduction, it is difficult to implement. This problem arises not only from the technical aspect, but also from organizational, communication, and psychological aspects. Moreover, improper reuse may cause unexpected cost and excessive waste of resource, which can be frustrating. Therefore, it is worthwhile to study these barriers and find ways to overcome them.

Technical difficulties mainly arise from the inherent nature of information and knowledge. In particular, design information is highly unstructured. Not only does the information consist of many facets, which cannot be represented in a consistent and exchangeable manner, but the same type of information can be represented in different formats. For example, many CAD modelers have their proprietary data structures for representing the geometric information of a product, which restricted the communication between them.

From the organizational perspective, design reuse is restricted by the traditional individual project-based management strategy. Companies usually do not provide incentives for engineers to engage in design reuse because such an effort does not directly add value to the project at hand. The reluctance of the managers to apply design reuse may be justified by the observation that if the time required to reuse a part is approximately greater than 30% of the time required to design the part from scratch, design reuse will fail [Girczyc and Carlson, 1993]. This observation is not necessarily applicable to all circumstances. However, it is reasonable for managers to detect evident benefits of design reuse before they would adopt such a methodology.

Communication can also be a big problem because individual designers may find it difficult to reuse the designs, experience, and insight of others [Busby, 1999]. This causes problems in the design reuse activities, including indexing, retrieval, and modification. For example, designers may have difficulties in finding the relevant designs to address new problems especially when the past designs have been developed in an environment different from the current one. Furthermore, the differences in the environment and the problem itself may make the understanding and modification of past design formidable.

Psychological problems happen when the designers feel no excitement or self-satisfaction to reuse other's ideas. Hence, some designers are reluctant to borrow existing ideas.

Other than these barriers, design reuse should not be applied for its own sake. In fact, excessive reuse can be harmful for a company. It can lead to design fixation and kill creativity if designers fail to seek solutions beyond the current reuse repository.

As an example, the authors of this book participated in a project aiming at applying the design reuse methodology to design an industry product, namely, the fan filter unit (FFU). Part of this project involved the design of the FFU casing. Figure 1.6 shows Solution A provided by the engineers using a design reuse method. The motor is fixed on a supporting plate, which is in turn underpinned by a reinforced plate. The reinforced plate is fixed on the casing walls using four nuts and bolts around four corners. Since this structure has been used and tested in many cases, it is very safe and well-documented. Only minor changes are needed to accommodate new design requirements. Consequently, the casing was designed quickly and efficiently, and the manufacturing and assembly is routine and cost effective. In comparison, Solution B was created by engineers without considering design reuse. In their design process, the traditional structure cannot provide the desired features. Therefore, they redesigned the casing structure such that the motor is supported by a bottom plate connected to the top cover plate using four long bolts. The entire supporting sub-assembly is, in turn suspended on the upper plane of the casing. This new design requires much more effort, time, and cost. However, the performance of the product improved significantly in terms of noise and vibration performance.

Without a proper design scenario and the design criteria, it is impossible to determine which solution is better. Nonetheless, this example is not aimed at comparing which solution is better. Rather, it highlights the point that design reuse is not omnipotent and that it should not be applied in all circumstances.

1.6 Summary

In this chapter, product design methodologies relevant to design reuse are discussed. Based on the discussions of the major issues in design reuse, it is evident that a comprehensive design reuse process model must accommodate the modeling, analysis and optimization requirements during the entire design process. Engineering design reuse applications are extensively studied with different disciplines, namely software engineering, mechanical and electro-mechanical engineering, and manufacturing. Furthermore, the barriers of design reuse are discussed to avoid possible pitfalls of reuse and 'overuse'.

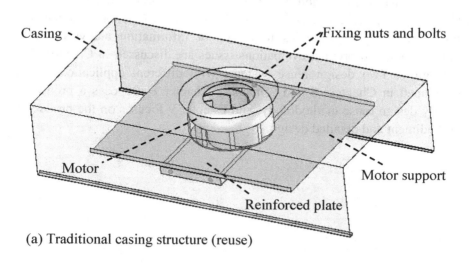

(a) Traditional casing structure (reuse)

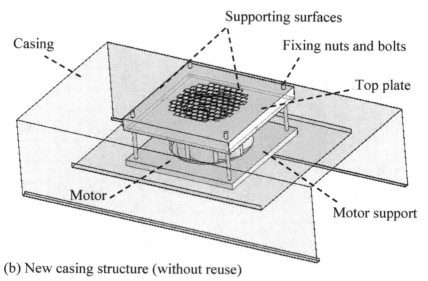

(b) New casing structure (without reuse)

Figure 1.6 FFU casing structure with/without design reuse

In the subsequent chapters, the techniques in design reuse are addressed progressively. Chapter 2 provides a literature review of the design reuse systems and the commonly used computational tools. Chapter 3 focuses on the modeling of product information. Chapter 4 investigates the technologies to carry out information analysis. The optimization and product evaluations issues are discussed in Chapters 5 through 7. Two design reuse systems with different applications are developed in Chapters 8 and 9, namely Chapter 8 proposes a product family design reuse methodology, and Chapter 9 focuses on the product embodiment and detailed design.

Chapter 2

Design Reuse Systems and Enabling Tools

This chapter investigates the legacy systems and techniques in design reuse. Many of these systems deal with the product design problem within the mechanical and electrical domain. The investigation includes the latest progresses of design reuse in the academia and the industry. Considering the imperative requirement of computation support in design reuse, Section 2.2 presents the prevailing computational tools in design reuse, which enable more efficient decision support. The power and applicability of these tools are discussed accordingly.

2.1 Engineering Design Reuse Approaches

Design reuse is an experience-oriented approach and has long been adopted by designers, consciously or sub-consciously. Although direct reuse of previous design components is not applicable in most design projects, it is a common practice for a designer to resort to similar past designs as a starting point, even in the original design. In the context of product development, a design reuse system must be supported by the domain knowledge in engineering design, computer technologies for the purpose of computation, as well as management sciences for optimum knowledge manipulation. This section presents a literature review of the notable design reuse approaches and systems.

2.1.1 *Case-based reasoning*

Case-based reasoning (CBR) is probably the most classical design reuse approach. In CBR, knowledge or experience is embedded in design cases, which are retrieved and reused with respect to similar design requirements. A typical CBR design cycle includes four stages: retrieval, repair, reuse and retain [Watson, 1999] (Figure 2.1). Among them, the retrieval stage is concerned with searching for the appropriate cases according to the design needs, usually using the nearest neighbor technique [Watson, 1999; Pearce *et al.*, 1992]. The repair stage, which is also called adaptation, adapts the retrieved cases to fit the new design requirements. Retrieval and repair are popular topics in CBR research.

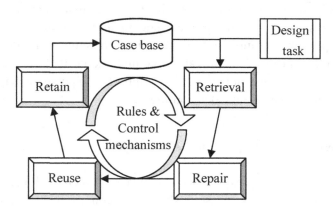

Figure 2.1 The case-based reasoning process model

CBR has been applied in many engineering domains. Several projects are discussed next to illustrate the possible applications. For additional information, readers can refer to a comprehensive review of CBR systems with an emphasis on the issues of case representation, retrieval, presentation and adaptation [Watson and Perera, 1997]. A number of useful links to the case-based applications is available at http://www.ai-cbr.org/projects.html.

- RODEO [Altmeyer et al., 1994; Altmeyer and Schurmann, 1996] was developed based on a formal model to describe design objects, design processes and the design requirements and specifications of their properties. It searches for the most suitable modules in a design database according to the requirements and specifications, and provides users with the necessary adaptation steps.
- CASECAD [Maher and Pu, 1997] defines a flexible indexing system. Hence, it provides designers with a browsing tool to navigate through design case histories or retrieve a specific design case using formalized specifications of new design problems.
- SPIDA [Manfaat et al., 1998] uses CBR and machine learning to retrieve and reuse a similar past design or generate an abstract layout. Pattern matching and information retrieval techniques are used to define case similarities and retrieve similar designs.
- HOMER [Göker and Roth-Berghofer, 1999] is a system that stores the expertise of the help-desk professionals in a case base and enables designers to access, reuse, and extend the knowledge in a natural and straightforward manner.

Despite these efforts, no individual CBR-based model covers all the significant aspects of design reuse and falls short as a comprehensive design reuse model. In conceptual design, CBR has not been very effective because it is more concerned with the selection and modification of instances rather than the generation and subsequent utilization of past design abstractions. Therefore, CBR is more suitable for variant design.

2.1.2 Catalog-based design

Catalog-based design is also known as component-based design. It focuses on the establishment of a component catalog that can be reused in future designs based on well-indexed catalog components. The components are usually derived from existing product cases, and are reused directly in new designs.

The component-based compositional method [Chakrabarti and Bligh, 1996] emphasizes the representation and abstraction of known structural components to generate solutions. It defines the design problem as an input-output structure. A set of mechanical components are abstracted as input-output elements with motion transmission and motion controls at each end. A few typical elements include tie rods, cams levers, screws, generic shafts, etc. This method involves a strategy to synthesize solutions based on an exhaustive search of matching mechanical elements.

A-Design combines multi-objective optimization with a multi-agent system for automated design synthesis [Campbell et al., 1999, 2000]. A number of electro-mechanical components are defined using function parameters (FPs). The system is capable of accepting changing design inputs and decision-making based on previous experience. Such capabilities make the system intelligent and adaptive to dynamic environments. However, one limitation of this system is its dependence on the FPs, which relies heavily on the input-output relations of the design variables, usually in the form of equations. This limits the flexibility to build the product architectures. Moreover, the design components in A-Design were hard-coded in a number of Lisp files making it difficult to update the relevant information.

The NIST design repository project is aimed at developing a computational framework to create design repositories that are accessible to distributed users [Szykman et al., 1998, 1999, 2001]. This research provides a means to manage the heterogeneous product information. It presents a generic product information model that is characterized by a formalized function-flow representation schema. It emphasizes the interoperability of the information model and the role of knowledge. However, it does not provide a comprehensive solution to design synthesis. Knowledge reuse is largely restricted to conventional CBR.

Overall, the catalog-based approaches lack a comprehensive process model to cover the multiple procedures in design reuse. Moreover, the catalogued components are more static than dynamic. It is difficult to develop a component catalog that can be easily adapted to meet the requirements of the dynamic design environment.

2.1.3 *Modular design*

As compared to catalog-based design, modular design is a more comprehensive method. In modular design, a set of building blocks, known as modules, is identified or created. A product family is derived by adding, removing or substituting one or a few modules to a base platform [Pahl and Beitz, 1996]. Modular design is closely related to the design reuse methodology. The modules are, in essence, elements that are designed to be reusable. Modular design usually involves the following processes: (1) the identification of a product architecture and reusable components (modules) from existing products, (2) the combination and adaptation of modules to generate new designs, and (3) the assessment of product cost and performance.

Modular product architecture has been discussed by Ulrich [1995], in which a product architecture is defined as the scheme by which the product functions are mapped to the physical components. Specifically, it involves "(1) the arrangement of functional elements; (2) the mapping from functional elements to physical components; (3) the specification of the interfaces among interacting physical components" [Ulrich, 1995]. A modular architecture is distinguished from the integral architecture in the way functional elements are mapped to the physical components. A modular architecture has a one-to-one mapping from the functional elements to the physical components of the product. Three basic types of modular architecture are defined, namely slot, bus and sectional, according to the interfaces between the components.

The development of modular product architectures has been a central topic. Reported researches include the function-based representation and quantitative methods [McAdams *et al.*, 1999, Stone *et al.*, 2000a, 2000b], the concept selection techniques [Mattson and Magleby, 2001], module identification and reuse [Allen and Carlson-Skalak, 1998], the design for variety (DFV) method [Martin and Ishii, 2002], etc.

Erixon [1996] proposed the design for modularity methodology where modularity was considered as the major factor for product and factory reuse and reengineering. The methodology employs a modular function deployment (MFD) technique to integrate five steps into a comprehensive framework. MFD makes use of multiple tools to ensure the modularity of products, such as QFD, Pugh's concept selection method, module indication matrix (MIM) questionnaire, interface/evaluation matrix, and design for manufacturability and assembly (DFMA).

Interface design is another important issue in modular design. Van Wie *et al.*, [2001] used the DSM tool to design interfaces in a modular product architecture, so as to reduce assembly costs. Blackenfelt and Sellgren [2000] developed robust interfaces for modular products. Sundgren [1999] dealt with the interface problem in the context of product families after extensive analysis of several Swedish manufacturing corporations.

Modular design has also found broad applications in the industry. A few representative ones are presented next.

Bally Engineering The Bally Engineering Structures employs a basic modular component, namely the pre-engineered panel to generate products in a variety of shapes and sizes. The system is especially responsive to the changing design requirements [Pine, 1993b].

Scania AB Modular Cab The Swedish heavy truck and bus maker Scania successfully applied the design for modularity method in the design of their truck cabs [Erixon, 1996]. The cabs are produced in the same production line, using a standardized assortment of modules and component. A number of product variants can be derived from eight basic types of cab, using reduced sheet metal parts, and with reduced changeover time.

Nippondenso Nippondenso Co. Ltd. provides customized components for various automobile manufacturers using modules with standardized interfaces. It is reported that 288 different types of panel meters can be made from 17 standard modules [Whitney, 1993].

Hewlett-Packard A number of models of the HP DeskJet and LaserJet printers can be developed using pre-defined modular components [Feitzinger and Lee, 1997]. The modular framework involves both modular product design and modular process design. This strategy has effectively enabled the postponement of product differentiation.

Modular design has also been applied in the design of PCs, *e.g.*, the IBM's system/360 [Baldwin and Clark, 2000], power tools, *e.g.*, Black & Decker [Lehnerd, 1987], watches, *e.g.*, Swatch [Ulrich and Eppinger, 2004], home electronics, *e.g.*, Sony Walkman [Uzumeri and Sanderson, 1995], automobiles, *e.g.*, the cockpit module of Volkswagen [Wilhelm, 1997] and the rolling chassis module of Dana [Kimberly, 1999], etc.

Despite the extensive studies in the academia and the successful stories in the industry, it should be noted that modular design is not necessarily suitable for all products. Some products may favor an integral architecture over a modular one. For example, the BMW R1100RS motorcycle employs an integral architecture based on function sharing, which allows for geometric nesting of components to reduce the space that a product occupies [Ulrich and Eppinger, 2004]. Ulrich [1995] discussed the factors to be considered in determining integral or modular architectures.

2.1.4 *Adaptable design*

Adaptable design (AD) is a new paradigm that addresses the challenges faced by contemporary design and manufacturing enterprises [Gu *et al.*, 2004]. Adaptability refers to the capability of products to extend their utility/service to address new requirements from the customers or producers, with business and environmental considerations. In this sense, the utility of a product can be extended because the product can evolve with the changing environment. Two types of adaptability are defined, namely product adaptability and design adaptability. The former refers to the ability of a product to accommodate various usages. The latter refers to the capability of a design (blueprint) to be modified to create new products.

AD brings benefits to both users and producers. For the users, they can enjoy extended product utility. For the producers, new products can be generated based on existing design plans, process plans, manufacturing setups, and existing parts and components. The benefits of AD to users and producers are more evident for large products because of the savings achieved through reuse and adaptation in large scale products and production systems is more significant [Gu *et al.*, 2004].

In line with AD, flexible design is a philosophy focusing on the products' adaptability/flexibility to the changing market [King and Sivaloganathan, 1998]. Flexible design adopts a reuse strategy that merges an existing design with other independent designs to develop a common 'core design'. Thus, a company can expand its business scope from its original sectors to other sectors with lower investments or resources. As compared to modular design, which usually operates within a single market sector, flexible design has better adaptability because it can help a company expand in both market niches and market sectors (Figure 2.2). For example, a cellular phone manufacturer may use a modular structure that is supported by a number of processors, displays, motors, switches, etc. The combinations of these modules generate product variants to address different market niches. In contrast, a company embracing flexible design may set its business in various digital entertainment devices, such as cellular phone, MP3, digital camera, portable media player, etc. These products can use a 'common core' that accommodates the requirements of multiple market sectors, as well as the market niches.

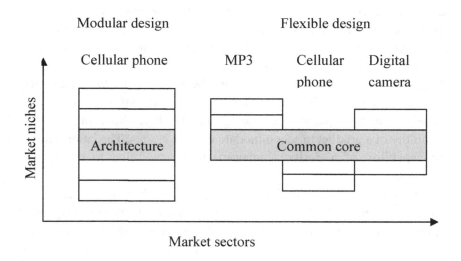

Figure 2.2 A comparison of modular design and flexible design

2.1.5 *Expert systems*

Expert systems were first developed by researchers in AI during the 1960s and 1970s, and were commercialized for various applications in the 1980s. It features a sizable body of researches that reuse expert intelligence and experience in tackling new problems. An expert system, also known as a knowledge-based system, is a computer program that contains some of the domain-specific knowledge of one or a few human experts [Rychener, 1989]. The knowledge can be processed to solve problems in the same way as the human experts do.

Expert systems have been developed due to requests from organizations that have a high-level of know-how experience and expertise that cannot be easily transferred to other members. Hence, an expert system contains not only the pieces of knowledge that are accumulated and articulated by an expert, but also the reasoning logics that direct the utilization of knowledge. A novice designer can learn the domain knowledge and logics as he/she interacts with the computer-based expert systems. These expert systems can be as simple as using true/false logic to evaluate data, or as complicated as performing sophisticated evaluation taking into account real-world conditions and uncertainties, using techniques such as fuzzy logic.

A typical expert system consists of four components, namely concepts, rules, models and strategies [Rychener, 1989].

Concepts refer to the declarative representation of domain objects, which can be abstract classes or concrete instances. These objects usually contain a set of major attributes based on which the similarities among them can be derived.

Rules are empirical associations and inferences, which involve cause and effect, evidence and hypothesis, etc. Rules are the basic reasoning elements to be applied in knowledge utilization.

Models refer to the collection and organization of interrelated rules. Models determine the processes of problem-solving and the interaction of the users and the system.

Strategies are the mechanisms to aid the usage of the knowledge base, such as guiding the search and resolving conflicts.

Research and applications of expert systems abound in literature. An extensive discussion of such systems would make the manuscript unnecessarily long. In light of the theme of this book, a collection of systems is analyzed with an emphasis on design reuse in terms of (1) objects modeled, (2) the representation scheme, (3) the reuse method, and (4) the knowledge update scheme (Table 2.1).

Table 2.1 Expert system and their support to design reuse

Systems	Objects modeled	Representation scheme	Reuse method	Knowledge update scheme
DESRU [MacCallum and Duffy, 1995]	design task process computational support	descriptive	reuse library synthesis using parts of the conceptual design information	Refinement and evaluation for implementation of software support
NODES [Duffy et al., 1996]	Generic conceptual objects, their qualitative and quantitative inter-dependencies	formal description method used in AI, OO method, abstraction with top-down product decomposition	Concept library for purpose of engineering design, configuration, cost estimation, strategic planning	Generalization
PERSPECT [Duffy and Kerr, 1993]	Design experiences	probably OO method; graph may used	multi-perspectives	Automated group rationalization
Scheme-builder [Counsell et al., 1999]	Design concept: function-means-component	OO method; hierarchical: function-means tree network	Schema library, Reasoning by expert system	General management operations (addition, deletion, modification)
CONGEN [Gorti and Sriram, 1996]	design concept: function-form-behavior, design rationale	OO method; Symbolic function; Form: shape, geometry, spatial relation	CBR; Function to symbol mapping; Symbol to form mapping	–

From the reuse perspective, an expert system is a useful mechanism to acquire and utilize human expertise. Like any computer-based systems, expert systems perform well in certain aspects, but show inadequacies in others. The advantages and disadvantages of an expert system are summarized in Table 2.2.

Table 2.2 Pros and cons of expert systems

Advantages	Disadvantages
Capture and maintain high levels of information	Some knowledge may not easily articulated and captured
Combine multiple human expert intelligence	The lack of human common sense needed in some decision-makings
Offer consistent support to repetitive decisions, processes and tasks	Inability in dealing with unusual circumstances
Improve efficiencies and reduce time needed to solve problems; reduce chances of human errors	Difficulty in automating complex processes
	The lack of flexibility and ability to adapt to changing environments
Facilitate systematic thinking of human	The lack of creativity in problem solving
Reduce training costs to novice members	
Review transactions that human experts may overlook	

2.1.6 *Innovative design using TRIZ*

TRIZ (or its equivalent TIPS) is a methodology, tool set, knowledge base, and model-based technology for generating innovative ideas and solutions for problem solving. It is developed by former Soviet engineer and researcher Altshuller [1984] based on his study of about 40,000 patents. Altshuller summarized 39 principles of invention. As a central theme, conflicts/contradictions are the key driving force of product invention. An inventive solution to a problem is often the one that overcomes some contradictions [Savransky, 2000]. From a functional point of view, a contradiction happens when a harmful function is generated with the change of one parameter to achieve a favorable function. A contradiction-driven process may be developed to generate concepts. It begins with a function model, by which some contradictions are identified. These contradictions are expressed with generalized parameters or engineering parameters that are related to specific physical effects of a product. Next, they are used in a TRIZ relationship matrix to seek generalized solution principles. From the process, the TRIZ methodology presents a systematic way to inventive problem solving.

TRIZ has been adopted as a powerful method to generate concepts and refine ideas, and has been extensively applied in the engineering domain. A number of commercial software packages have been developed to reduce the time needed to solve innovative problems, *e.g.*, Goldfire Innovator (http://www.invention-machine.com/), TRIZSoft from Ideation International Inc. (http://www.ideationtriz.com/software.asp), Guided Innovation Toolkit (GIT) from Pretium Consulting Services (http://www.pretiumllc.com/Soft.htm), CREAX Innovation Suite from CREAX of Belgium (http://www.creaxinnovationsuite.com/), TRIZ.it! Innovation Principles (http://www.triz.it/), etc. These software packages also provide vast case libraries that can be visited and reused.

Despite its sound theoretical foundation and impressive structure, TRIZ is by no means a replacement of human intelligence. The theory and tools can be used to guide the thinking and reasoning of humans. However, creative thinking is still a gift of humans. Human intelligence is such a complex system that, as yet no man-made systems can completely simulate it, nor supersede it.

2.2 Reasoning in Design Reuse

Advancements in computer science and software development have aided design reuse. A significant amount of studies has been carried out to use computers for knowledge processing and reuse. Typically, a design reuse system makes use of one or a few AI techniques to automate the reasoning processes to a certain extent. This section investigates the widely adopted AI techniques that support design reuse.

2.2.1 *Machine learning*

As its name indicates, machine learning is concerned with the development of techniques that allow computers to 'learn'. In a narrower sense, machine learning refers to inductive machine learning, which focuses on the creation of computer programs to extract rules and patterns out of the massive data sets [Potter *et al.*, 2001]. A systematic approach to machine learning can be developed based on the transformation of existing knowledge into new knowledge. A generic model of such a process is shown in Figure 2.3, together with the elements of learning. Among these elements, the knowledge transformer is the core. Based on the nature of design knowledge and the learning goals, seven types of knowledge transformers have been identified [Sim and Duffy, 1998].

- Group rationalization (grouping)/ decomposition (ungrouping)
- Similarity/dissimilarity comparison
- Association/disassociation
- Derivation/randomization
- Generalization/specialization
- Abstraction/detailing
- Explanation/discovery

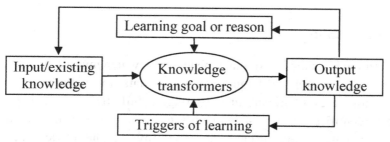

Figure 2.3 Processes and elements of learning [Sim and Duffy, 1998]

Machine learning was implemented as a generalization process in NODES [Duffy and MacCallum, 1996]. Whenever a new case is added to the library, the system compares its associated knowledge with the knowledge in the existing cases, and makes generalizations. The generalization process is carried out recursively so that the addition of new concepts or cases is reflected in the updated knowledge base. In PERSPECT [Duffy and Kerr, 1993], a knowledge base generator is used to perform automated group rationalization to group product instances according to their major features. An attribute graph method is used to organize the outcome of the grouping. The group rationalization method was also adopted in BRIDGER, which is a design synthesis system for bridges [Reich, 1993]. Design cases are modeled as a hierarchical structure based on taxonomies in BRIDGER. Machine learning was in the form of a toolbox program in 'n-dim', which is a multi-dimensional information modeling and reuse framework [Reich *et al.*, 1993]. The toolbox is capable of natural language processing and produces a terminology tree structure to reveal the inherent patterns of design information.

Machine learning is a major enabler of knowledge extraction for design reuse. Existing systems have focused mainly on the modeling languages, the representation of learned information, and the identification of patterns from unstructured data. Despite these efforts, more extensive work is required on the usability and validity of the knowledge learned from raw data, namely: (1) the methods to present and share knowledge to multiple and distributed users, and (2) the methods to ensure that the learning output is valid in new situations.

2.2.2 *Data mining*

Data mining is a sub-field of machine learning, which involves the process of automatic search of large volumes of data for patterns, such as association rules [Adriaans and Zanting, 1996]. It is concerned with pattern identification, which is much the same as in machine learning. However, data mining usually does not involve the process of rule extraction as in machine learning. Basically, data mining can be divided into two types based on the nature of data to be analyzed.

Structured Data Mining
- database mining
- graph mining
- sequence mining
- tree mining
- web mining
- concept mining

Unstructured Data Mining
- text mining
- image mining

Successful applications of data mining mainly lie in business, management, and science. Recent years also witnessed the increasing applications of data mining in engineering.

The rough set theory was used to deduce rules from the feature data of products [Kusiak and Tseng, 2000]. This data mining method was used to solve the selection problems in engineering design, namely the prediction problem in the printed circuit boards. The rough set theory has also been applied in many engineering problems in the industry [Berry and Linoff, 1997]. CLEMENTINE enables users to extract knowledge that is explainable with respect to the qualitative model [Clark and Matwin, 1993]. This system was applied to two process control systems, namely water tank network and ore grinding process. An association rule mining system (ARMS) was proposed for effective production portfolio identification. ARMS differentiates the customer needs from the functional requirements, and accordingly, the product portfolio can be identified based on the mapping relationship between the customer domain and the functional domain. This method can improve the efficiency and quality of product portfolio identification by alleviating the tedious, ambiguous and error-prone process.

Similar to machine learning, data mining is a powerful tool for knowledge extraction. However, its application depends on the effective interpretation, utilization and validation of extracted knowledge.

2.2.3 *Design structure matrix*

A design structure matrix (DSM) is a tool for modeling the constituent sub-systems/activities and identifying the corresponding information exchange and dependency patterns among them. The core of DSM is a matrix representation of design activities, which is a derivative of directed graph that shows the interdependency and sequence relationships of the constituent elements. DSM was first developed by Warfield [1973] and

Steward [1981] and has been recognized and widely adopted as a useful tool for system design since the 1990s.

DSM can be used to model and represent product or project development data, and accordingly the data can be analyzed based on matrix operations, such as clustering and sequencing. Thus, design patterns can be recognized, which is helpful to reconstruct the system structure. Based on the data type it deals with, DSM can be divided into four types (MIT DSM web site: http://www.dsmweb.org/).

Component-based DSM is used to model multi-component relationships. A typical data analysis method is clustering.

Team-based DSM is used to capture the characteristics of multi-team interfaces, and is usually adopted in organization design. Clustering is the major data analysis tool.

Activity-based DSM is mainly useful for project scheduling and activity sequencing. Activity input/output relationships are captured and analyzed using sequencing and partitioning operations.

Parameter-based DSM models parameter decision points and necessary procedures. It is mainly used in low level activity sequencing and process reconstruction.

Research and applications of DSM abound in engineering design. Scott and Sen [1998] proposed to use DSM to sequence the design activities and incorporate the potential of concurrency in a project environment, where the reuse of design data is facilitated. The project emphasizes the decision support for design reuse initiatives by incorporating evaluation criteria, such as the design lead time. Eppinger *et al.* [1994] used a matrix representation to capture both the sequence and the technical relationships among many design tasks. These relationships define the 'technical structure' of a project, which is analyzed to find alternative sequences and/or definitions of the tasks. Such improved design procedures offer opportunities to speed up the development progress by streamlining the inter-task coordination. One limitation of these applications is that the numerical form of the elements in a DSM is not easily dealt with by human. Moreover, this activity is difficult to carry out until the detailed design stage. Kusiak and Wang [1995] developed an algorithm to derive the dependencies between the design variables and the goals. A fuzzy-logic-based approach is used to model imprecise dependencies between variables in the case when no sufficient quantitative information is available. Dong and Whitney [2001] made efforts for early-stage capturing of system interactions, *i.e.*, analyzing the system before the detailed stage. The strategy is to resort to a design matrix based on axiomatic design and transform it into a DSM using pre-defined matrix operations. However, this method did not prove how well the matrix transformation could predict the system interactions. To compensate for this limitation, Hommes and Whitney [2003] developed detailed procedures to validate the completeness of matrix transformation.

Many prototype and commercial DSM software are available, such as
- PSM32 (http://www.problematics.com/)
- PlanWeaver (http://www.planweaver.com/)
- Lattix (http://www.lattix.com/)
- Excel Macros for Partitioning and Simulation (http://www.dsmweb.org/dsm_tools/DSM-Program.xls)
- MATLAB Macro for Clustering DSMs (http://www.dsmweb.org/)

A limitation of DSM is that the construction of a DSM requires significant effort and expertise. Currently, this is carried out mainly through reading documentation and interviewing the engineers and managers who are working on the system of interest [Dong and Whitney, 2001]. Such a practice makes the process subjective and inefficient.

2.2.4 *Artificial neural networks*

Artificial neural network (ANN), or commonly known as neural network (NN), is an information processing tool composed of a collection of neurons that are organized in a mathematical model. A typical ANN is used to model complex, non-linear relationships between input and output (regression), or to find patterns in the input data (pattern classification). Two fundamental issues for constructing a network are: (1) designing the network architecture, and (2) training the network to perform well based on the training data. The basic elements of a network architecture are the neurons assigned to one or more layers. For example, Figure 2.4 shows a two-layer network structure where the input is denoted as a vector $\vec{x} = [x_1, x_2, ..., x_n]^T$. Each output neuron is connected to the input neurons via a weight vector w^i. The output can be computed as a function of the summation of weighted input, $y_j = \varphi(\eta_j)$, where φ is called the activation function. Training of a network involves the determination of the proper values of the weights in the architecture. This is usually achieved through back-propagation [Rumelhart et al., 1994].

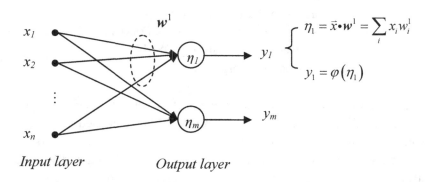

Figure 2.4 Neural network structure

There are two types of training, namely supervised and unsupervised training. The former requires a training data set consisting of the input data and the expected output (the target). The output generated by the network may deviate from the target. Training is carried out by varying the weights such that the deviation is minimized. In contrast, the training data does not contain the output information in an unsupervised training. Hence, the system is expected to evolve into an equilibrium status based on pre-defined criteria.

Applications of ANN techniques have achieved significant progress in design and manufacturing. Su [1998] used ANN in retrieving design knowledge and comparing it with expert systems. According to the results obtained from the design of power transmission system, ANN has the advantages of: (1) adaptability to changes in the design information, and (2) ability to deal with incomplete data. However, it suffers from prolonged training time and difficulty in interpreting the input/output data. Potter et al. [2001] used ANN to acquire heuristic design knowledge in the design of fluid power circuits. The ANN was able to learn certain simple associations between the input attributes and the solution elements. However, the performance of the network was not satisfactory when there is insufficient training data. Kamarthi and Kumara [1993] applied the NN technique in concept generation. As a central element, an Idea Generation System (IGS) is introduced that organizes existing product information to support decision-making. In particular, multi-layer perceptron and adaptive resonance theory (ART) networks have been used to address the classification and mapping problems. Li et al. [2006] proposed a hybrid method that combines the ART2 network, heuristic reasoning, and graph manipulation to recognize interacting manufacturing features. The method outperforms traditional approaches, such as the rule-based approach, graph-based approach, hint-based approach, etc., in that it can recognize more types of features with enhanced adaptability to new features.

Despite these applications, ANN in design and manufacturing suffers a few deficiencies.

(1) The performance of the system depends much on the availability and validity of the training data. However, typically in engineering design, there is no guarantee that sufficient and valid data can be obtained.

(2) The output data pattern may be difficult to interpret. In other words, even though the ANN system generates attractive design patterns, the meaning and utility of the outcome relies a lot on the experience of the engineers, which may restrict the utilization of the information.

2.2.5 Genetic algorithms

Genetic algorithm (GA) encompasses a number of optimization algorithms motivated by the natural evolution process [Goldberg, 1989]. During a GA-based optimization, a population of potential solutions (called the individuals) is maintained. These individuals are evaluated according to the fitness functions, *i.e.*, their fitness values are computed according to the fitness functions. Next, the individuals compete with each other in the reproduction process based on their fitness values. Thus, an individual with a larger fitness value has a better change to reproduce. Two basic types of reproduction are carried out
- **Crossover** – Individuals, called parents, are paired and their structures are decomposed according to specific rules. The offsprings are generated by recombining the pieces of the parents.
- **Mutation** – Part of the structure of the offsprings is modified at a certain probability (mutation rate).

Subsequently, the offsprings are re-evaluated and re-inserted into the population. These processes are reiterated until a pre-defined termination criterion is satisfied or no substantial improvement can be made to the objectives. A flowchart of GA is shown in Figure 2.5.

An effective application of GA must address the following issues: (1) the representation of the problem (the chromosome structure), (2) the reproduction scheme (3) the definition of the fitness function, and (4) the control parameters of the GA process, *e.g.*, population size, crossover/mutation rate, termination criteria, etc. These issues will determine the efficiency of the computation and the effectiveness of the design space exploration.

The charm of GA lies in its power in solving non-linear, multi-modal optimization problems, where traditional random search methods fall short of inefficient and the linear programming methods may face difficulties in finding the global optima. Hence, GA is especially suitable for solving large combinatorial problems, which are typical in design synthesis problems. The applications of GA in engineering design will be further studied in Chapter 5. A GA-based method is proposed to solve the product configuration design problem.

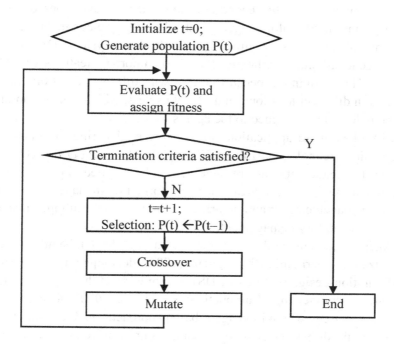

Figure 2.5 A flowchart of genetic algorithm

2.2.6 *Agent-based method*

An agent, which is sometimes referred to as an intelligent agent, is a component of software and/or hardware that is capable of performing certain tasks according to the user's demand. Agents are characterized by three attributes [Jennings, 1995]:

- **Autonomy** (pro-activeness) refers to the ability of agents to operate on their own without human guidance.
- **Social ability** (cooperation) means that agents are capable of interacting with other agents and contribute to the common objective.
- **Reactivity** (learning) means that agents are able to perceive their environment and improve their performance with time and experience.

An implication of the second attribute is that an agent-based system usually consists of multiple agents instead of a single agent. Thus, agent systems are, in a sense, equivalent to multi-agent systems (MAS). The multi-agent structure further enables two important features of MAS, namely: (1) distribution, because agents with different functions can be situated in different locations, and (2) robustness, which is made possible by the redundancy designed to the agents.

MAS has broad applications in software engineering, business, and information technologies [Liu and Zhong, 1999, Zhong et al., 2001]. In the past decade, it has been extensively applied in the design manufacturing domain, in areas such as information management, supply chain management, manufacturing planning, scheduling, material handling, inventory management, etc.

Agent-based method has been used in design synthesis and optimization. Darr and Birmingham [1994] developed the automated configuration-design service (ACDS) system to meet the requirements of concurrent engineering. Campbell et al. [1999, 2000] developed the A-Design system, which is an agent-based approach capable of automated design synthesis based on catalogue components. For example, Moss et al. [2004] proposed to use agents to enable a basic learning mechanism that allows for the transfer of knowledge across design problems. The authors augmented the A-Design system using a mechanism to break down (chunking) the existing designs and store chunks to be utilized to solve new problems. Agent-based methods have been used to facilitate scheduling in manufacturing. For example, Zhou et al. [2003] proposed a hybrid hierarchical MAS model for agile job scheduling in a virtual workshop environment. Three layers are defined in this model, namely the scheduling manager agent layer, task agent layer and production resource agent layer. Using this architecture, the scheduling system is simpler and its reliability and robustness are improved.

The agent-based method gives the designers more flexibility and control over the designed system. However, this technology should not be overused. In essence, the power of the system relies much on the processing capacity of the agents and a proper coordination among the agents. Being a variant of the AI technologies, the agent-based method faces the same fundamental problems that AI technologies face, such as effective representation of knowledge, modeling unstructured information, creativity in design, etc.

2.3 Summary

This chapter presents the state-of-the-art technologies in design reuse. The discussions follow two themes, namely the legacy design reuse systems and the enabling tools for reasoning in design reuse. Six types of systems have been studied. The discussion shows the basic principles of each system, followed by an investigation of the working models in the academia and the industry. Reasoning in design reuse has been supported by a vast pool of tools and technologies. This chapter discusses a few most notable ones. It should be noted that these AI techniques are not mutually exclusive. Some common elements may exist amongst these techniques in different forms. Moreover, some systems can be based on a hybrid of more than one AI technique.

Chapter 3

Product Information Modeling

Product information takes various forms and is subject to changes. Therefore, a comprehensive information model is required to capture the multiple facets of product information. Moreover, a proper representation scheme is necessary for the exchange and reuse of information, which is becoming more and more evident with the increasing collaboration across distributed design teams. A formal representation of product information has been advocated to accommodate these requirements. This chapter provides an overview of the fundamental modeling techniques of product information. A function-based model is developed which incorporates the multiple facets of the product information.

3.1 Data, Information and Knowledge

Design reuse emphasizes the reuse of knowledge to facilitate decision-making. However, knowledge may not be immediately available from the product cases. It is derived from raw data, and grows gradually towards a more comprehensive structure. Design reuse has to consider the 'rawness' level of the knowledge. Bergmann [2002] considered *knowledge* in relation to and as distinguished from *data* and *information*. The distinctions and associations between these fundamental concepts are summarized in Figure 3.1.

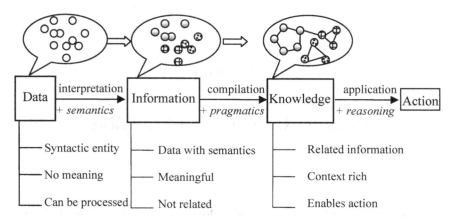

Figure 3.1 Relationships between data, information and knowledge [Bergmann, 2002]

Data is a syntactic entity and is meaningless by itself. It can be processed manually or using computers. Data becomes meaningful after interpretation, and accordingly, is considered as information. For example, '1500 watt' and 'automatic power off' are data, which are meaningless when considered in isolation. However, in the context of the technical specifications of an electric kettle, namely the power consumption and power control, they become meaningful.

Information is data with semantics and meaning. It is very important in problem solving. However, information is not immediately useful in directing an action to solve a problem. For example, the power consumption, '1500 watt', of an electric kettle does not entail what type of heating devices that an electric kettle may have to use. Moreover, the pieces of information are not necessarily related to each other.

Knowledge is a set of interrelated information through a compilation process. In product design tasks, knowledge actuates actions to achieve the desired goals. Problem solving is achieved through a reasoning process where knowledge is applied and a set of rules can be extracted. For example, when designing an electric kettle, the power consumption, the size and shape of the product, the product life expectancy, and cost, collectively determines the type of heating device and the technical parameters of the heating device. The designer takes the responsibility to decide on the final solution based on available knowledge.

In the design reuse methodology, the starting point is a bundle of data and information of the existing product cases. Often, a profusion of raw data is available from an enterprise. This chapter focuses on the modeling of product information such that the bundle of data and information can be identified. Thereafter, effort is needed to find the relevant information and discover the underlying patterns so that it becomes informative. This is the work of knowledge extraction, which will be discussed in the subsequent chapters.

3.2 Information Modeling – State-Of-The-Art Review

This section focuses on the formal representation of product information with an emphasis on four important issues, namely (1) the *content of the information model* which determines the type of information that should be included; (2) the *modeling language* is a major enabler of information exchangeability and interoperability; (3) *taxonomies* define the basic vocabulary that constitutes the modeling language; finally, product data should be supported by (4) *database systems in a web-based environment* for better maintenance efficiency.

3.2.1 Content of information model

The product information model must contain the basic ingredients to be reused in future designs. Therefore, this model must be comprehensive enough to capture the multiple facets of the product design data. From the viewpoint of a designer, a product may involve multiple information facets that are useful to decision-making (Figure 3.2).

3.2.1.1 Form

Traditional CAD/CAM systems focus mainly on the representation of the geometric data and geometry-related information, such as constraints, parametric information, and features [Szykman *et al.*, 2001]. These are typical information belonging to the 'form' (also called structure) category. During the product development process, the form information is usually determined at the detailed design stage. Since design freedom is low at this stage, *i.e.,* the designer could only make minor changes to the design, the reusability of the form information is insignificant. Hence, traditional CAD systems are inadequate in providing comprehensive solutions to systematic product development based on reuse.

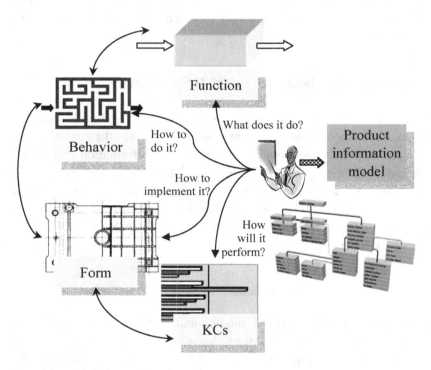

Figure 3.2 Multiple facets of product information

3.2.1.2 Function

Among the spectrum of product information, function has been recognized as a critical element. Hence, a function-based information model has been widely adopted in literature. Typically, function is considered as the purpose or intended use of a feature, component, or product [Ulrich and Seering, 1987; Baxter *et al.*, 1994]. More specifically, function is considered as a general relationship between the input and the output of a system with the objective of fulfilling a task [Pahl and Beitz, 1996]. For a complex technical system, a hierarchical function structure is built to demonstrate the conversion of flows (energy, material and signal) between functions (Figure 3.3). This input-output view has been adopted and extended by many researchers [Gero, 1990; Gorti and Sriram, 1996; Kirschman and Fadel, 1998; Szykman *et al.*, 1999; Otto and Wood, 2001].

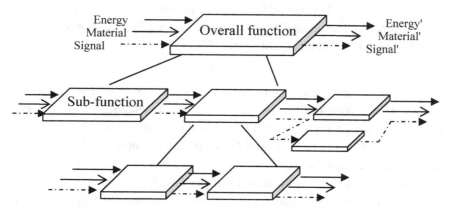

Figure 3.3 Function structure with flow conversions [Pahl and Beitz, 1996]

Stone and Wood [2000] proposed a function basis for product design. The function basis has been motivated by several factors, such as the product architecture design, the storage and transmission of information, creativity in concept generation, etc. It enhances representation rigor and reasoning logic. Similarly, the NIST design repository project presents a generic product information model that is characterized by a formalized function-flow representation schema [Szykman *et al.*, 1998]. Effort was also made to reconcile the above two approaches [Hirtz *et al.*, 2002].

In value analysis, a function is considered as any task that is to be fulfilled by means of an object, a process or an activity. Whichever viewpoint is adopted, the consensus is that a function is an abstraction from the physical artifacts, and hence, is not dependent on specific implementation. Thus, function-based design allows for greater creativity.

3.2.1.3 Behavior

Function alone cannot accommodate the multiple facets of product information. Typical in the AI field is the combination of behavior with function to allow for better decision-making [Umeda *et al.*, 1990; Chandrasekaran *et al.*, 1993; Iwasaki *et al.*, 1995]. Behavior refers to the underlying principles or processes that make the related function attainable, *i.e.*, it describes how a system behaves to fulfill the desired function. The simplest format of behavior is probably the equations based on physical laws. For example, the energy (E) that is generated by a heating device of an electric kettle can be derived from the properties of the resistance (R) and electrical current (I).

$$E = I^2 R \tag{3.1}$$

In a more generic manner, behavior can be expressed as a Causal Process Description (CPD) [Iwasaki and Chandrasekaran, 1992]. A CPD is defined as a pair $\{C, G\}$, where C is the condition under which the device is expected to behave, and G is a directed graph $G=\{N, L\}$. N is a set of nodes, each of which represents a partial description of a state. In particular, N_{init} represents the initial state and N_{fin} represents the final state. L is a set of directed links among the nodes. Thus, the CPD represents the transition of a system from one state to another under certain conditions.

The behavior of a function is context sensitive and is evident only when the form is determined to a certain extent. As such, behavior is considered as a link between function and form. For example, the function of converting electrical energy to heat exists in both an electric kettle and a microwave oven. However, it is implemented based on different principles. Accordingly, the physical laws to describe the behaviors of the function are different.

Many researchers have made efforts to combine the multiple facets of production information. Thus, function-form-behavior models have been extensively studied [Iwasaki and Chandrasekaran, 1992; Gorti and Sriram, 1996; Qian and Gero, 1996; Szykman *et al.*, 2001; Roy *et al.*, 2001]. This strategy allows product information to be modeled in a comprehensive way. Different facets have different usage and support product design from different aspects. Thus, it is very useful in design support.

3.2.1.4 Key characteristics

Key characteristics (KCs) are typical properties of a product in aspects such as performance, appearance, quality, reliability, cost, etc. They can be qualitative or quantitative. For example, Table 3.1 lists a subset of KCs of a diesel engine.

Table 3.1 Key characteristics of a diesel engine

Fuel type		Diesel	
Cylinder type		Vertical, single cylinder	
Combustion system		Direct injection	
Displacement (cc)		212	
Engine speed (rpm)		2800	3500
Max.output (hp)		3.8	4.2
Starting system		Recoil/Electric	
Power take off		Crankshaft or Camshaft	
Fuel tank capacity(L)		2.4	
Electric system		12V-18AH	
Net weight(kg)		26.5	
Dimensions(mm)	L	325	
	W	405	
	H	419	

Although KCs are not a systematic description of a product design, they can capture information at various abstraction levels and from multiple perspectives. Moreover, they facilitate the efficient interaction between suppliers and customers through the identification and use of critical information. The concept of KCs is a main enabler of knowledge-based product development [Rezayat, 2000].

3.2.1.5 Representation of product family

The above representation schemes have focused more on individual products. Recently, there is a trend towards the representation of a product family. Generic bill-of-material (GBOM) was used to explore the generic product architecture and identify the assembly structure of a product family [Erens *et al.*, 1994]. As an extension of GBOM, Generic Product Modeling (GPM) was used to represent product families from the business and assembly viewpoints [McKay *et al.*, 1996]. The functional information has been overlooked in GBOM and GPM. A Product Family Classification Tree (PFCT) was developed to model product configuration knowledge from the functional viewpoint [Yu and MacCallum, 1995]. However, the interrelations between the modules and the end-products are not explicitly included in PFCT. A generic Product Family Architecture (PFA) was proposed, which explicitly deals with functional, behavioral and structural information [Jiao and Tseng, 1999]. The Programmed Attribute Graph Grammar (PAGG) was developed based on the graph grammar to specify the design space and assist product family generation [Du *et al.*, 2002a]. Furthermore, a graph rewriting program was developed which enables the derivation of product variants through graph transformations [Du *et al*, 2002b]. The graph grammar-based modeling is excellent in formal, visual, and extensible product family representations. However, the generation of a product family based on a few graph transformation operators falls short of being too restrictive. This new method requires further development to deal with more complex product family design problems.

3.2.2 *Modeling languages*

Modeling languages allow product information to be represented consistently and concisely. The product models in various CAD systems usually adopt vender-specific file format, which is highly proprietary. At present, most CAD systems provide interface and automatic conversion to exchange the CAD models between different systems, and to convert a CAD model to neutral formats, such as STL (Stereolithography), IGES (Initial Graphics Exchange Specification), and VRML (Virtual Reality Modeling Language). However, it leaves much to be desired due to the loss of parameters, constraint, and features during the conversion process [Pratt and Anderson, 2001]. This makes the transferred model difficult to be reused and revised in the downstream receiving system. Hence, the vender-specific file formats are not suitable for information exchange.

Several generic modeling languages have been employed in engineering design, such as UML [Pulm and Lindemann, 2001; Felfernig *et al.*, 2001], CML [Bobrow *et al.*, 1996], STEP [Pratt and Anderson, 2001; Szykman *et al.*, 1998], EXPRESS [Kahn *et al.*, 2001], and XML [Szykman *et al.*, 1999; Rezayat, 2000]. These languages provide a common syntax with well-defined semantics to model a broad variety of physical processes and objects. They are also favorable for the exchangeability, accessibility and interoperability of the product information between different design groups. Table 3.2 summarizes the key features of the prevalent generic modeling languages.

Table 3.2 Generic language in information modeling

Language	Features	Application area
Compositional Modeling Language (CML) [Bobrow *et al.*, 1996]	Supports model sharing between different research groups, and enhances interchange and reuse of domain theory; Provides common syntax with a well-defined semantics.	A wide variety of physical phenomena, process and objects.
Extensible Markup Language (XML) Developed by the World Wide Web Consortium (W3C)	Provides a text-based means to implement a tree-based structure to information modeling. A non-proprietary neural format and is designed to be extensible by allowing user-defined elements based on Document Type Definition (DTD). Because of its structured format and platform-independent feature, XML is suitable to the sharing of data across different systems, particularly systems connected via the internet.	XML is not restricted to certain applications. It has been extended to various domains such as geography, mathematics, archival, music, etc.
Integration DEFinition Language 0 (IDEF0) released by the National Institute of Standards and Technology (NIST)	IDEF0 (also referred to as Integration Definition for Function Modeling) is a method designed to model the decisions, actions, and activities of an organization or system. IDEF0 is a functional modeling language building on SADT™ (Structured Analysis and Design Technique™). IDEF0 is useful in establishing the scope of an analysis, especially for a functional analysis to promote good communication between the analyst and the customer.	Functions (activities, actions, processes, operations) related to multiple domains.

Table 3.2 Generic language in information modeling (continue)

Language	Features	Application area
Standard for the Exchange of Product model data (STEP) (ISO 10303) Developed by the International Organization for Standardization.	The ISO 10303 (STEP) consists of a set of application protocols (APs) to enable the exchange of product-related data between different CAD systems or between downstream engineering systems. It captures product information from multiple aspects, such as product description and support, geometrical and topological representation, product structure configuration, parameterization and constrains, etc.	A wide range of product types: mechanical, electronic, electro-mechanical, sheet metal, process plant, architectural. Various PLC stages: planning, design, analysis, manufacturing.
EXPRESS (ISO 10303-11) Developed by the International Organization for Standardization.	EXPRESS is the data modeling language of STEP. It is an object-oriented language and has a scheme for creating a hierarchy of entity types. An EXPRESS model can be textual or graphical.	Generic objects of various types.
Unified Modeling Language™ (UML™) Developed by Object Management Group (OMG)	UML is a general-purpose modeling language that is an industry standard for specifying software-intensive systems. UML 2.0, the current version, supports thirteen different diagram techniques, which can be divided into three categories, namely structure, behavior and interaction (Ambler, 2004). UML is extendable and offers a mechanism for customization.	General-purpose modeling language. Software-intensive systems.

Information modeling based on generic language is not without problem. A general criticism is that these languages are very large and complex, making the learning curve unreasonably long. Moreover, information processing based on these languages is not straightforward due to the high level of abstraction. In product design, it is equally important to focus on the products, instead of the 'entities' or 'physical phenomena' that are highly abstract [Bobrow et al., 1996].

3.2.3 Taxonomies

The need for a standardized taxonomy in engineering design is critical for the following reasons.
(1) Information modeling is ambiguous [Szykman *et al.*, 1998]. Designers often use different terms to mean the same thing, or use the same term to mean different things, which creates a great barrier to the efficient retrieval and reuse of knowledge.
(2) Computer-based reasoning about natural language is still rare. Due to the large amount of vocabulary and pervasive subtle meanings of natural language, processing of information using computational algorithms is impractical at present. To make use of the power of computers in information processing, it is necessary to develop a set of terminologies that are both concise and precise.

Therefore, the purpose of using taxonomy is to represent the product information with a limited number of vocabularies, which are "as small as possible, yet generic enough to allow modeling of a broad variety of engineering artifacts" [Szykman *et al.*, 1999]. Current progresses in taxonomy development are mainly made in the function-based fields [Altshuller, 1984; Hundal, 1990; Pahl and Beithz, 1996; Kirschman and Fadel, 1998; Szykman *et al.*, 1999; Otto and Wood, 2001]. A few most notable ones are discussed next.

Table 3.3 Basic function groups [Kirschman and Fadel, 1998]

Motion	Rotary, Linear, Oscillatory, Other
	Create, Convert, Modify, Dissipate, Transmit
	Flexible, Rigid
Control	Power, Motion, Information
	Continuous, Discreet
	Modification, Indication
	User-supplied, Internal Feedback
Power/matter	Store, intake, Expel, Modify, Transmit, Dissipate
	Electrical, Mechanical, Other
Enclose	Cover, View, Protect
	Removable, Permanent
	Support, Attach, Connect, Guide, Limit

Pahl and Beitz [1996] divided the input/output flows into energy, material and signal, and developed a set of generally valid functions in the form of verbs (*e.g.*, change, vary, connect, etc.). Otto and Wood [2001] refined Pahl and Beitz's taxonomy with eight categories of functions (channel, support, connect, branch, provision, control magnitude, covert, signal) and three categories of flows (energy, material and signal). However, the taxonomy includes various 'synonyms' and 'compliments', which makes it complicated to be computationally processed. Kirschman and Fadel [1998] classified the vocabularies into four groups, namely motion, power/matter, control and enclosure (Table 3.3). They were further arranged in a sentence form leading to approximately 150 combinations of elementary mechanical functions. However, the sentence form is not suitable for developing a rigorous function model. The taxonomy adopted by Szykman *et al.* [1999] consists of six function categories and three flow categories (Figure 3.4). The categories adopt a multi-level structure that has dual effects. On the one hand, it makes the taxonomy generic and flexible, and on the other hand, it presents difficulties for computational processing.

Function
    Usage-function [...]
        Sink [absorb, consume...]
        Source [add, create...]
        Storage [accumulate, collect...]
    Combination/distribution-function [...]
    Conveyance-function [...]
    Signal/Control-function [...]
        Mathematical/Logical [...]
    Assembly-function [...]

Flow
    Material [...]
        Solid [...]
        Liquid [...]
        Gas [...]
        Multi-phase-mixture [...]
    Energy [...]
        Generic [...]
        Mechanical-domain [...]
            Translational-domain [...]
            Rotational-domain [...]
        Electrical-domain [...]
        Thermal-domain [...]
        Hydraulic-domain [...]
    Signal [...]

(a) Function Taxonomy   **(b) Flow taxonomy**

Figure 3.4 Function and flow taxonomy [Szykman *et al.*, 1998]

3.2.4 *Database system and web-based environment*

In the development of a comprehensive product information model, a distributed and collaborative design and manufacturing environment should not be overlooked (Figure 3.5). There is an increasing awareness of the sharing and exchange of information within and across enterprises. At the same time, the World-Wide-Web (WWW) is becoming a powerful and versatile information repository. Some even considered it as the largest knowledge-based system ever built [Fensel *et al.*, 1997]. However, the immense capability and versatility of the Internet also brings a side-effect, namely the effective management of product information via the Internet is not straightforward due to the following reasons.

(1) The amount of information on the Internet is immense, making the retrieval of relevant information a daunting task.
(2) Product information appears in diverse forms, such as texts, graphics, hypertext, CAD models, and video or audio sequences. The indexing and retrieval of these diversified data formats is not easy.
(3) Existing database systems are not especially tailored to the requirement of product information modeling. There is no pre-defined structure that allows information providers to represent information in a way that it can be efficiently retrieved and reused.

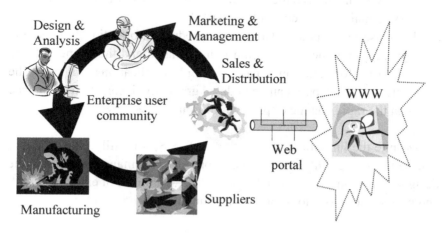

Figure 3.5 The enterprise community and the web-based environment

In this section, the characteristics of tradition Database Management Systems (DBMS) and product database management (PDM) systems are discussed, which lead to the requirements of information modeling with database support in a web-based environment.

3.2.4.1 Traditional DBMS

DBMS based on relational database is an established technology. Many commercial DBMS are available to provide secured, persistent and integrated solutions at different scales (*e.g.*, Oracle®, Microsoft® SQL Server, IBM® DB2, MySQL®, etc.). However, in light of product data management, traditional DBMS are inadequate in a number of aspects.

(1) Traditional database systems are string-based, and hence more efficient in processing textual and numerical strings than graphical and unstructured data. As discussed earlier, product information appears in diverse forms and is highly unstructured, which renders existing DBMS inflexible in data storage and management.

(2) DBMS are more data-centric than knowledge-centric [Szykman *et al.*, 1998]. A relational database consists of a number of records which relationships are pre-defined by indices and stored procedures. However, the underlying logic among the pieces of information is not easy to maintain based solely on DBMS.

(3) Database systems are static while product information is dynamic. Typically, product information is interrelated and the change of one piece of information may cause multiple implications. Although the content of a database can be updated gradually, it cannot keep in pace with the rapid change of product information.

Despite the limitations of traditional DBMS, it is still a powerful and mature technology that can be used in product data management. It is the designer's responsibility to make use of the capabilities of existing DBMS and provide solutions to information maintenance and reuse.

3.2.4.2 PDM systems

PDM is an established technology that aims to create an automatic link between product data and a database. It is process-oriented and focuses on the downstream information communication among and/or across different design departments. As collaboration is becoming prevalent in today's manufacturing environment, PDM is a major enabler to facilitate collaborative product design.

The information being stored and managed include engineering data, such as CAD models, drawings and their associated documents, product visualization data, and various metadata, such as the owner of a file and the release status of the components. The major functionalities of a PDM include:

- Control of check-in and check-out of the product data to multiple users;
- Management of engineering change and release control on all versions/issues of components in a product;
- Building and manipulation of the product structure bill-of-material (BOM) for assemblies;
- Assistance in configurations management of product variants; and
- Corporation-wise management of complex products to spread product data over the entire product lifecycle management (PLM) process.

However, there is an appeal for the PDM systems to include function, behavior and structure information [Bilgic and Rock, 1997], which is more related to the front-end product design.

Based on the above discussions, product information modeling has to consider the following aspects with database support in a web-based environment.

- The database system must be able to capture the multiple facets of product information, such as geometry, function, behavior, and associated files.
- The relationships among the information chunks must be properly maintained when changes are made to the product data.
- The designer must develop mechanisms within or beyond the database system to provide multiple views to different users.

- A scalable database system is preferable to accommodate the different requirements of the enterprises at different scales.

3.3 Function-Based Product Information Model

This section presents a product information model developed by the authors as part of their effort to develop a product family design reuse (PFDR) system. The product information model is used to capture the multiple facets of product information, as is embedded in the existing product cases. The model provides the representation rigor and flexibility to allow for subsequent analysis and reuse.

3.3.1 A multiple facet product information model

A product case p_i is represented as a 4-tuple, and denoted as:

$$p_i \sim \left(F_i^o, K_i^o, M_i^o, X_i^o\right) \qquad (3.2)$$

where
(1) F_i^o denotes a hierarchical function structure, which is obtained through function decomposition. This book adopts the Function Analysis System Technique (FAST) method to carry out the function decomposition process [Otto and Wood, 2001]. The decomposition stops when each of the functions can be fulfilled by a single, basic solution principle. Such a function is called an **atomic function**. Thus, a flow-oriented, hierarchical function structure is established for each product case. Two types of relationships may exist between the functions, namely descendant and communication. The former refers to whether a function has one or a few sub-functions (children). The latter describes two functions that are connected by flow(s): a function can either be the source or the destination of another function according to the direction of the flow(s). Figure 3.6 shows the data structure of function and flow.

(2) $K_i^0 = [k_{i1}, k_{i2}, ..., k_{ip}]^T$ is a vector of the KCs that signify the performance features of p_i. They are qualitative or quantitative engineering specifications that are related to the customer requirements. The data type of a KC can be one of the following: categorical, ordinal, Boolean, and real number, as shown in Figure 3.7. Among them, the real number is continuous while the other three are discrete. The categorical data represents a set of mutually exclusive values. For example, the material of a part can be 'stainless steel', 'aluminum', 'plastic', etc. A Boolean type refers to whether a characteristic is present or not, and the value of a Boolean type KC is restricted exclusively to 'True' and 'False'. An ordinal type is similar to the categorical type except that it has an ordering feature, *i.e.*, a higher order value can satisfy a lower order requirement. For example, the service cleanliness of a clean room can have the following grades {100000, 10000, 1000, 100, 10, 1}. The succeeding ones can satisfy the preceding ones but not *vice versa*.

(3) $M_i^0 = [m_{i1}, m_{i2}, ..., m_{in}]^T$ contains the physical components that implement the corresponding product functions, as shown in Figure 3.8. Each physical component (m_{ij}) implements one or a few atomic functions. The physical component is basically a geometric model, preferably a CAD model. The attributes of the physical component are included, such as material, dimension, cost, weight, etc.

(4) X_i^0 represents the contextual information. It captures the relationships between the different information facets, namely (1) the relationships between KCs and functions (denoted as X_{K-F}), which describe whether a KC is dependent on one or a few functions, and (2) the interdependency relationships between different physical components (denoted as X_{M-M}), which refer to whether two components are compatible with each other. Figure 3.9 illustrates this.

Each product case is analyzed and modeled according to the data structures. Similar products are assigned to a virtual design space (SP_i). Among the multiple facets, the functional facet is supported by the formal function representation scheme and taxonomies discussed next.

```
TFunction  // TYPE: function
{
    name: String;        //name of the function
    input: TFlow;        //input flow
    output: TFlow;       //output flow
    action: String;      //a verbalization of function
    sub: TFunction;      //subordinate/child function block
}
TFlow  // TYPE: flow
{
    type: Enumeration;   //type of flow ($E_{NG}$, $M_{AT}$, $S_{IN}$)
    name: String;        //unique name of a flow
    source: TFunction;   //function block as the source of flow
    dest: TFunction;     //function block as the destination of flow
}
```

Figure 3.6 Data structure of function and flow

```
TKC  // TYPE: Key characteristics
{   name: String;
    type: Enumeration;  //4 types: categorical, Boolean, ordinal, real
    value: Variant;
}
```

Figure 3.7 Data structure of KCs

```
TPhysical    // TYPE: physical (geometric model)
{   name: String;
    ref_fun: TFunction;      // related function.
    att: TAttribute;         //attributes of the physical component.
    ref_model: String;       //reference to the physical model.
    modeler: Enumeration;    // CAD modeler.
}
```

Figure 3.8 Data structure of physical components

```
TContextual   // TYPE: Contextual
{   name: String;
    kc_f: TMap;     // mapping relation between KC and function.
    m_m: TCom;      // compatibility relation between physical components.
}
```

Figure 3.9 Data structure of contextual information

3.3.2 Representation of function using key element vector

Based on the flow-oriented, hierarchical function structure, an atomic function is defined in terms of the input/output flows and the function actions, *i.e.*, the input flows are transformed into the output flows as a result of the action of the function [Pahl and Beitz, 1996]. A flow can be one of the three forms, namely energy (E_{NG}), material (M_{AT}) and signal (S_{IN}), which are defined in the flow taxonomy. An action of a function is in the form of a transitive verb that is defined in the function taxonomy. Based on these definitions, each atomic function is described as follows, where [] denotes optional object(s):

transitive verb(s) + [the input flow] + [the output flow]

For example, an electric kettle must have a function of heat generation. A 'heat generation' function may involve 'converting' the input energy 'electric wattage' to the output energy 'heat'. As a result, the function can be represented as: *convert electric wattage to heat*. Figure 3.10 shows a block representation of the function.

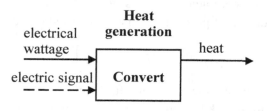

Figure 3.10 A block representation of function - 'heat generation'

Thus, an atomic function is formally represented as three main attributes, namely the input flow (I_W), output flow (O_W) and the function action (A_F). Each function action, in combination with the corresponding input/output flows, constitutes a key element κ. Thus, a key element is represented as a vector:

$$\kappa = \begin{bmatrix} A_F & I_W & O_W \end{bmatrix} \quad (3.3)$$

where I_W is the input flow(s), and $I_W \in E_{NG} \cup M_{AT} \cup S_{IN}$, and O_W is the output flow(s), and $O_W \in E_{NG} \cup M_{AT} \cup S_{IN}$.

If a function involves multiple function actions and multiple key elements, this function is represented as a key element vector (KEV), denoted as $\bar{f}$:

$$\bar{f} = \begin{bmatrix} \kappa_1 & \cdots & \kappa_m \end{bmatrix}^T = \begin{bmatrix} A_{F1} & I_{W1} & O_{W1} \\ \cdots & \cdots & \cdots \\ A_{Fm} & I_{Wm} & O_{Wm} \end{bmatrix} \quad (3.4)$$

where m is the total number of function actions. Usually m should not be a large value, as in such a case, the function can be further decomposed into sub-functions, e.g., the 'heat generation' function can be represented as a one-element KEV: $\bar{f} = \begin{bmatrix} \text{convert} & \text{wattage:electric signal} & \text{heat} \end{bmatrix}$.

Finally, assuming that the characteristics of non-atomic functions can be defined by the complete set of its descendant atomic functions, the function structure F_i^0 of product p_i can be converted into the KEV format:

$$F_i = \begin{bmatrix} \bar{f}_1 & \bar{f}_2 & \cdots & \bar{f}_M \end{bmatrix}^T \quad (3.5)$$

where M is the total number of atomic functions.

The elements in a KEV are non-numeric, and as such they are not suitable for computational analysis. The solution to this problem involves two steps: firstly, developing a taxonomy with proper coding schemes; and secondly, mapping the function actions and flows to quantitative values using these coding schemes.

3.3.3 *Function taxonomies*

The taxonomy in this book is extracted and refined from the existing research. In particular, the flow taxonomy is defined for three main types, namely energy, material, and signal, with one level of sub-category in each type. The function action taxonomy includes four basic types. Three types are defined with respect to the three flow types; and the last type deals with the assembly and spatial relations of product components, known as the 'enclosure' functions. Each of the first three categories involves one level of sub-category. In comparison to the reported work, the taxonomy developed in this book has a simpler structure, such that they can be easily coded. Table 3.4 and Table 3.5 shows the function and flow taxonomy, respectively.

The taxonomy is further coded such that a function action is represented by a unique 3-digit code and a flow is represented by a unique 4-dight code (Figure 3.11).

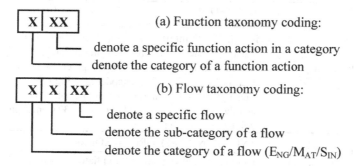

Figure 3.11 Coding schemes of function action and flow taxonomies

Using these coding schemes, each function action and each flow in the taxonomy is assigned a unique code. The function action taxonomy code spans from 000 to 999, and the flow code, 0000 to 9999. The codes are predefined in the system such that a user will use them to build the function models using simple 'click and assemble' operations. For example, the code of the 'heat generation' function is

$$\vec{f} = \begin{bmatrix} 302 & 1202:3005 & 1304 \end{bmatrix}.$$

Table 3.4 Function taxonomy

Category		Function Action
Energy and Material	Usage [1]	absorb(01), consume(02), destroy(03), dissipate(04), eliminate(05), empty(06), export(07), remove(08)
		add(21), create(22), emit(23), supply(24), extract(25), generate(26), import(27), provide(28)
		accumulate(41), collect(42), store(43)
	Combination/ Distribution [2]	combine(01), connect(02), couple(03), link(04), mix(05), branch(10), distribute(11), divide(12), separate(13), sort(14)
	Transformation [3]	attenuate(01), convert(02), filter(03), modify(04), refine(05), amplify(11), increase(12), decrease(21)
	Conveyance [4]	advance(01), channel(02), conduct(03), convey(04), direct(05), divert(06), guide(07), move(08), rotate(09), transfer(10), translate(11), transmit(12), transport(13)
Signal	Generation [5]	generate(01), open(02), turn-on(03), emit(04), store-value(05), display(06)
	Processing [6]	adjust(01), decrease(02), delay(03), detect(04), display(05), equalize(06), enhance(07), increase(08), inhibit(09), limit(10), maintain(11), measure(12), resist(13), select(14), sense(15), amplify(16), demodulate(17), attenuate(18), compare(19), decode(20), decrypt(21), digitize(22), encode(23), filter(24), interrupt(25), modulate(26), reset(27), split(28), switch(29), toggle(30), track(31), vary(32), encrypt(33), isolate(34), time(35),
	Logical/ Mathematical [7]	AND(01), NOT(02), OR(03), XOR(04)
		add(11), decrement(12), differentiate(13), divide(14), increment(15), integrate(16), invert(17), multiply(18), shift(19), sort(20), subtract(21)
	Elimination [8]	turn-off(01), filtrate(02), close(03)
Enclosure [9]		assemble(01), constrain(02), cover(03), disassemble(04), enclose(05), extract(06), fasten(07), fix(08), guide(09), join(10), link(11), locate(12), orient(13), position(14), release(15), remove(16), secure(17), separate(18), stabilize(19), support(20), unfasten(21)

Table 3.5 Flow taxonomy

Category	Domain	Flow
Energy	Mechanical [11]	friction(01), gravitation(02), centrifugal force(03), contact(04), inertia(05), momentum(06), torque(07), human force(08)
		rotary motion(51), angular displacement(52), angular velocity(53), angular acceleration(54); translation motion(60) position(61), displacement(62), velocity(63), acceleration(64) translation(65); oscillatory(70), combinational(80)
	Electrical [12]	charge(01), wattage(02), electromotive force(03), current(04) voltage impulse(05), electrical impedance (06), resistance(07) capacitance(08), inductance(09)
	Thermal/ Chemical [13]	entropy(01), temperature(02), entropy flow(03), heat(04) combustion(05), oxidation(06), combustible gas-(1307)
	Hydraulic [14]	pressure(01), flow(02), volume(03)
	Optical [15]	reflection(01), refraction(02), diffraction(03), interference(04), polarization(05), infra-red(06), visible(07), ultra violet(08)
Material	Solid [20]	rigid body(01), elastic body(02), widget(03), powder(04), particulate(05), granular-matter(06), composite material(07), aggregate material(08)
	Liquid [21]	incompressible liquid(01), water(02), compressible liquid(03) homogeneous-liquid(04), petrel (05), diesel (06)
	Gas [22]	homogeneous(01), inhomogeneous(02), air(03), oxygen(04) nitrogen(05), carbon dioxide/CO_2(06), compressible(07), incompressible (08), flammable gas(2209)
	Multi-phase mixture [23]	solid-liquid(01), liquid-gas(02), liquid-particle(03)
Signal	Single [30]	sine wave(01), unit step(02), sinusoid(03), impulse (04), electric signal (05), switch on (06)
	Status [31]	sound(01), temperature(02), pressure(03), verbal(04), tone(05), visual(06), position(07), displacement(08), smell(09)

The taxonomy scheme used in this book can support a broad variety of product functions. However, the taxonomy is not necessarily complete as some functions may not have been defined in the taxonomy. Two methods are used to solve this problem. Firstly, the system accepts user-defined functions to allow for a certain level of ambiguity. Secondly, the taxonomy can be updated to include new vocabularies that are constantly being used. Another limitation is that the taxonomy scheme is not strictly orthogonal, *i.e.*, the function models are not free of inconsistencies because different designers may use different vocabularies to describe the same function. To alleviate this problem, the function modeling is supported by Graphical User Interface (GUI) with ample help text and explanation, such as the pop-up descriptions of a taxonomy and examples to show its usage.

3.3.4 *An illustrative example*

The product information of an electric kettle is captured and modeled using the aforementioned model. Details are presented next.

Function The function structure of an electric kettle was established through function decomposition. This function structure features a four-level hierarchy and seven atomic functions (Figure 3.12). The atomic functions of this product are shown in Table 3.6. Accordingly, Table 3.7 lists the input vectors that are retrieved from the coded KEV representation of atomic functions. The coding scheme follows the taxonomies presented in Table 3.4 and Table 3.5.

Physical The CAD model (in SolidWorks®) of this product is shown in Figure 3.13. All the geometry information and spatial relationships between the components are available from the CAD model. A tree-view assembly structure is shown on the left. The major components and sub-assemblies are displayed on the right, respectively.

Contextual Two types of contextual relations are considered in the product information model, namely (1) the relationship between the KCs and product functions, which is useful to determine the parameters that influence the product performance; and (2) the compatibility between the physical components. The first type of relationship is represented as a correlation matrix, as shown in Table 3.9. In this table, '1' indicates that a KC is dependent on a function; and '0' otherwise. The second type of relationship can be represented in diverse forms, such as spatial relationship between components in the CAD model, and textual and/or graphical descriptions. For example, the diameter of the heating disk must not exceed that of the base plate (Figure 3.14). Thus, a heating disk and a base plate are not compatible with each other if their dimensions do satisfy the above relation.

KCs Seven KCs of the electric kettle are identified (Table 3.8).

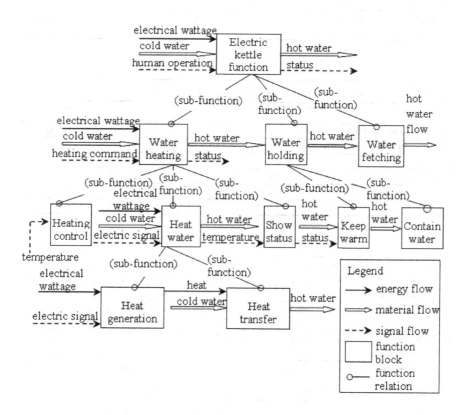

Figure 3.12 Function structure of an electric kettle

Table 3.6 Atomic functions of four sample products

NAME	A_F	I_w-E_{NG}	I_w-M_{AT}	I_w-S_{NG}	O_w-E_{NG}	O_w-M_{AT}	O_w-S_{IN}
keep warm	store		water			water	
heat transfer	conduct	heat	water			water	
water fetching	convey	human force	water			water	
heat generation	convert	wattage		switch on	heat		
show status	display			temperature			visual
contain water	enclose		water				
heating control	turn-on			temperature			switch on

Table 3.7 Normalized input data of the atomic functions

NAME	A_F	I_w-E_{NG}	I_w-M_{AT}	I_w-S_{NG}	O_w-E_{NG}	O_w-M_{AT}	O_w-S_{IN}
water fetching	404	1108	2102	0	0	2102	0
heating control	503	0	0	3102	0	0	3006
show status	506	0	0	3102	0	0	3106
heat generation	302	1202	0	3006	1304	0	0
heat transfer	403	1304	2102	0	0	2102	0
keep warm	143	0	2102	0	0	2102	0
contain water	905	0	2102	0	0	0	0

Table 3.8 KCs of the electric kettle

KCs		Unit	Type	Value	Description
k_1	Power consumption	watt	real	1000	A major determinant of the time required to heat up the water.
k_2	Dimension	mm	real	355*156*165	Dimension is specified by (height*width*depth)
k_3	Water capacity	liter	real	1.75	The maximum volume of the container.
k_4	Automatic control	–	Boolean	Yes	Automatic power cut-off when the boiling point is reached.
k_5	Water fetching method	–	categorical	Air pressure	A method of how the user can get water out of the container.
k_6	Cost	S$	real	35	Product cost.
k_7	Mean Time Between Failures (MTBF)	hour	real	15,000	A measure of the reliability of the product.

This section presents an example of the multiple facets of a simple electric kettle. It should be noted that given a particular product, a designer may not be able to collect each and every information facet. In such a situation, the designer may have to resort to some assisting tools or techniques, such as reverse engineering, to collect product data.

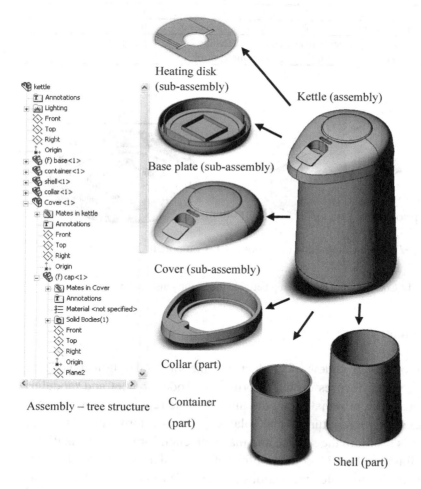

Figure 3.13 CAD model of an electric kettle

Table 3.9 Relation between KCs and atomic functions

KCs		f_1	f_2	f_3	f_4	f_5	f_6	f_7
k_1	power consumption	0	1	0	1	0	0	0
k_2	dimension	0	0	0	0	0	1	0
k_3	water capacity	0	0	0	0	0	1	0
k_4	automatic control	1	0	0	0	0	0	1
k_5	cost	1	1	1	1	1	1	1
k_6	MTBF	1	1	1	1	1	1	1
k_7	water fetching	0	0	1	0	0	0	0

f_1: Keep warm f_2: Heat transfer f_3: Water fetching f_4: Heat generation
f_5: show status f_6: contain water f_7: Heating control

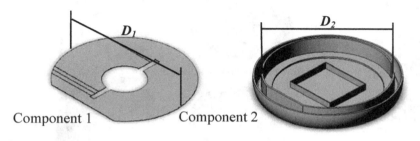

If $D_1 > D_2$, then Compatible (Component 1, Component 2) = False

Figure 3.14 Compatibility relationship between the heating disk and base plate

3.4 Summary

A state-of-the-art review of product information modeling techniques is studied, which focuses on the information model content and modeling languages. The information model must provide representation rigor and the flexibility in capturing the volatile product information. Database support is useful to the storage and management of product information.

A function-based product information model is proposed which effectively incorporates the various aspects of product data. This model highlights the following features:

- Multiple facets of product information are dealt with. The KCs can signify the performance features of a product. The function is an abstraction of product utility in the form of the relationship between input and output. The physical information provides a visual and editable information entity. The contextual information defines the relationship between multiple information facets.
- The function model is quantified using a KEV scheme.
- Comprehensive function and flow taxonomies are developed to enhance representation rigor.

Chapter 4

Design of Product Platform

Design reuse is closely related to product platform as a product platform is an effective means to organize existing product information that can be utilized in new designs. This chapter focuses on platform-based product development and the establishment of product platforms in the context of design reuse. In such an effort, (1) a computational method using neural networks is proposed to establish a function-based product architecture (FPA), and (2) a set of information processing technique is proposed to extract knowledge from existing products. This product platform is used as a basis in the design-by-reuse methodology.

4.1 Role of Product Platform

From a systematic perspective, product development involves a series of mapping processes throughout the entire spectrum of product realization, which encompasses four domains, *viz.*, the customer domain, functional domain, physical domain and process domain [Suh, 2001]. Product platform is a linchpin in the platform-based product development processes, and is responsible for establishing the technology core to correlate the functional requirements (FRs) with the design parameters (DPs). Figure 4.1 illustrates the position of product platform in this scenario.

The definition of product platform varies according to different perspectives and scopes. A prevailing definition was given by Meyer and Lehnerd [1997] where a product platform is considered as "a set of subsystem and interfaces developed to form a common structure from which a stream of derivative products can be efficiently developed and produced". This definition highlights three important features of a product platform, *viz.*, (1) a common structure, which means that a platform must be shared by a group of to-be-designed products, (2) sub-systems and interfaces, which are the content of a product platform, and (3) efficiency in developing product variants, which is the desired benefit of a product platform. Based on this definition, the major concern is to exploit the technology core that enables the logic of platform commonality and product variant differentiation. Knowledge must be well-organized within the framework of product platform to provide decision support.

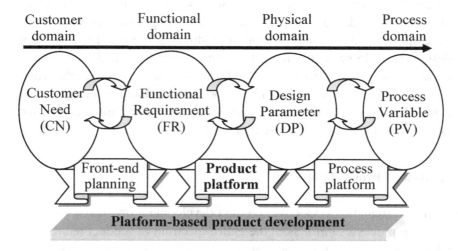

Figure 4.1 The role of product platform in the spectrum of product development

The core of the product platform is the product architecture, which is synonymous to the topology/layout of product functions and their embodiment. A product architecture refers to the scheme by which the product functions are mapped to the physical components [Ulrich, 1995]. The product architecture can be either integral or modular. Modular architecture has been repeatedly advocated because of its efficiency and flexibility in the management of complex systems. For example, different functions are delegated to different modules and product variants can be generated based on different combinations of the modules. This has been applied in mass customization based on an assemble-to-order rationale.

To summarize, this section explains three interconnected issues, *viz.*, platform-based product development, product platform, and product architecture. These issues have different implications and play different roles in product development. Their relationships are illustrated in Figure 4.2.

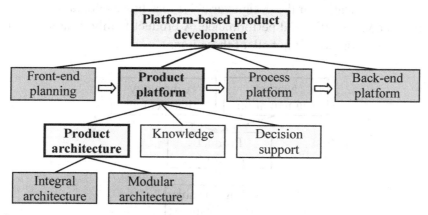

Figure 4.2 Platform-based product development, product platform and product architecture

4.2 Product Platform and Product Family Design

The purpose of a product platform is to launch a group of product variants which form a product family. From the perspective of product platform development and utilization, product family design can be divided into two basic types, *viz.*, (1) scale-based product family design, and (2) module-based product family design [Simpson *et al.*, 2001]. The former features a top-down perspective which emphasizes the strategic planning and design of the product platform and product family. Product variants are generated by 'stretching' or 'shrinking' the product platform with respect to a set of scaling variables. The latter features a bottom-up perspective which depends on the analysis and reuse of products and product components. A modular product architecture is constantly being used to organize the components and facilitate product development.

4.2.1 *A top-down perspective*

Figure 4.3 shows a top-down perspective in product family design. A product platform is developed based on market analysis and technology advancement. Next, product variants are generated by varying the DPs to achieve the desired functionality. Decisions have to be made concerning the division of the market segmentations, the determination of the design specifications, the choice of the variables to control product performance, and the optimization of the design variables to achieve optimal trade-offs between commonality and performance. A product family design system has to deal with most, if not all, of these issues.

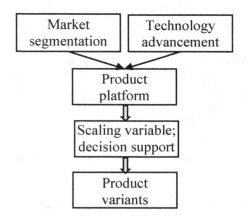

Figure 4.3 A process of top-down product family design

Among these efforts, the market segmentation grid was articulated, and the product leveraging strategies were proposed to utilize the sharing logic and cohesive architecture [Meyer and Lehnerd, 1997]. A robust concept exploration method (RCEM) was proposed to build a robust product platform that can accommodate a wide range of customer requirements [Chen *et al.*, 1996]. However, this is only the first step of a product family design. A second step, in which products are instantiated based on the platform, is equally important. A product platform concept exploration method (PPCEM) was proposed to support scale-based product family design [Simpson *et al.*, 2001]. This method explicitly defines two stages, *viz.,* product platform design and scale-based product family design based on the platform. However, PPCEM has two limitations. First, the commonality of the product family is determined by the designers based on a trial-and-error process. Second, the commonality is defined at only one level. In order to deal with the first limitation, a variant-based platform design methodology (VBPDM) was proposed to determine the design variables that should be made common among the products [Nayak *et al.*, 2002]. For the second limitation, a hierarchical platform design method was proposed to accommodate multiple levels of commonality in the product family [Hernandez *et al.*, 2002].

The methodologies in the top-down approach are effective only when a product platform can be properly defined. However, the information required to build the product platform is immense because the dimensionality of the design space is usually high. The dimensionality of the design space refers to the number of design variables, constraints, and objectives that have to be considered in a problem. A designer has to spend a lot of time and effort to study the intrinsic relationships between the product characteristics and the various design variables. Since relevant information may not be available, decisions may have to be made without proper context, possibly leading to sub-optimal solutions. For example, several top-down approaches have been applied to design the universal electric motors [Meyer and Lehnerd, 1997; Simpson *et al.*, 2001; Nayak *et al.*, 2002; Hernandez *et al.*, 2002]. Different strategies have been adopted to choose the design variables, set the constant and varying variables, and set the feasible ranges of the variables. Accordingly, different configurations of product family have been produced, which may not necessarily be compatible with each other. It is difficult to determine the configuration that would lead to the best design practice.

4.2.2 *A bottom-up perspective*

A bottom-up perspective leads to the module-based approaches. The product platform can be established through an analysis of the existing products. Based on this product platform, new products can be developed

using various design synthesis tools. Module-based design usually involves the following processes:
(1) Identification of product architecture and reusable components (modules) from existing products,
(2) Combination and adaptation of modules to generate new designs, and
(3) Assessment of product cost and performance. A modular design system should cover most, if not all these processes.

The design processes are illustrated in Figure 4.4.

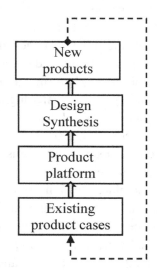

Figure 4.4 A process of bottom-up product family design

Module-based approaches are applied based on a set of existing products. Since the modules and product architecture are partially known, more information is available as compared to the scale-based approaches. As such, information deficiency can be alleviated provided that the information of the existing products can be effectively identified. However, module-based approaches have been criticized for their reliance on a large number of existing products [Hernandez et al., 2002]. Design freedom may be reduced if existing technologies are improperly utilized. Therefore, it is worthwhile to assess the reusability of existing products such that the design components can be logically reused and product quality ensured.

4.3 Computational Tools for Product Architecture Building

In the module-based approaches, the modularity of the product architecture is the key issue. The building of a modular product architecture starts from the decomposition of the individual products. Basic components are identified from the functional and/or structural viewpoints. Next, the components are analyzed such that modularity and commonality are identified, which culminate in the clustering of the components into logical modules. Previously, these activities have been carried out mainly by human engineers based on their experience. The efficiency and repeatability of such a practice is poor. Therefore, computational tools can be used to facilitate this process. This section introduces a few computational tools to support the building of product architecture. Their strengths and weaknesses are discussed accordingly.

4.3.1 *QFD-based approach*

Quality Function Deployment (QFD) defines procedural steps to find robust design concepts that address the customer requirements, which involve qualitative and quantitative evaluation of the interrelationships between customer requirements, technical requirements, and product planning. There is an extensive use of matrix operations to establish the relationships in the form of a house of quality (HOQ). On the other hand, the product architecture usually involves the mappings between the functional domain and the physical domain. There is an apparent similarity between them. Hence, it is natural to resort to QFD and its derivatives to facilitate product architecture design.

An important concept in product architecture is modularity. Identification of modularity in product design has constantly been represented as matrix operations. Huang and Kusiak [1998] proposed an approach to represent modularity and develop generic modular products. Gu and Sosale [1999] developed a modularization method to enhance modularity from different design perspectives. A similar rationale was adopted in the framework of the House of Modular Enhancement

(HOME) [Sand et al., 2002]. However, these methods were aimed at individual products instead of a product family.

QFD was used to identify the right design specifications [Erlandsson et al., 1992]. The QFD matrix (i.e., HOQ) thus created enables module creation, interface analysis, and modular configuration. As an extension, the QFD method was developed into the modular function deployment (MFD) method to evaluate module integration [Erixon and Ostgren, 1993]. MFD is similar to QFD, but the modularity drivers, instead of customer requirements, are mapped against functions. In the design for modularity method [Erixon, 1996], QFD is used to clarify product design specifications, and a QFD-like tool called the module indication matrix (MIM) is used to indicate the sub-functional groups that can form a module. QFD and MIM establish technical solutions to meet the requirements of product modularity. Martin and Ishii [2002] proposed the DFV method to build modular product architecture for multiple generations of products where a QFD-based, two-stage process was used to create the generational variety index, which is a measure for the amount of redesign effort required for future designs.

The limitation of the QFD-based methods is that the values within the QFD matrix, and the structures alike, are assigned by the engineers. Thus, the processes depend a lot on the designers' knowledge and experience. The method does not provide much computational support, and the repeatability of the process is poor.

4.3.2 *DSM-based approach*

Design Structure Matrix (DSM) was originally invented to organize product development tasks or teams to minimize unnecessary design iterations, and thus help to manage and speed up the development process. However, it can also be used to determine product modularity [Dong and Whitney, 2001; Yu et al., 2003; Hölttä-Otto, 2005]. A DSM can embody the interactions of functions/components, and algorithms can be designed to operate on the matrix to group functions or components. The objective is to maximize the interactions within groups and minimize interactions between groups. Figure 4.5 shows a simple scenario of using DSM to cluster components into modules. In Figure 4.5, seven components are involved in the design. Initially, the relationships among these components are unclear. By rearranging the components, three modules can be identified, *viz.*, module one (1, 2, 5), module two (3, 7) and module three (4, 6).

	1	2	3	4	5	6	7
1	x	x			x		
2	x	x			x		
3			x				x
4				x		x	
5	x	x			x		
6				x		x	
7			x				x

Original status

	1	2	5	3	7	4	6
1	x	x	x				
2	x	x					
5		x	x				
3				X	x		
7				X	x		
4						x	x
6						x	x

Grouped status (modules shaded)

Figure 4.5 DSM for module identification

Fixson [2002] used DSM to identify modular and integral architecture through the analysis of the number of functions that a physical component may provide in relation to other components. Yu *et al.* [2003] applied the DSM to identify highly coupled groups of product elements and to cluster them into logical modules. The authors adopted GA and the principle of minimum description length (MDL) to enhance the computation efficiency. Hölttä and Salonen [2003] compared the DSM method with the MFD method by applying them in various commercial products. It is claimed that DSM can simplify the module interactions, and hence it is most suitable for modularizing complex systems where interactions are too many to be handled manually. In addition, DSM outperforms MFD in operation repeatability. Hölttä-Otto [2005] proposed that DSM is especially suitable for quick rearranging of the architecture based on the interface interactions. However, the method is inadequate to deal with business oriented factors and product functionality. It depends much on the designer's judgment after the initial simplification of the architecture.

4.3.3 Heuristic and quantitative approaches

One limitation of the above methods is that they have not employed a formal representation of the product information. A formal representation of product functions enhances the establishment of product architecture. Quantitative methods have been proposed to represent product function, and identify product architecture based on functional interdependence and product similarity [McAdams et al., 1999; Stone et al., 2000a].

The function structure heuristic method is based on the function structures reported by Pahl and Beitz [1996], which is a decomposition block diagram of the functions of products represented as material, energy, and information flows. Since function is an abstraction of product structures, the function structure of different products is expected to share a certain kind of similarity. Based on this assumption, product architecture can be identified by invoking the innate similarities of the function structures of similar products. Thus, heuristic methods have been proposed to identify the architecture and modules based on a set of heuristic rules [Zamirowski and Otto, 1999; Stone et al., 2000b]. Most notably, three sets of heuristics have been developed, viz., (1) dominant flow heuristic, (2) branch flow heuristic, and (3) conversion-transmission heuristics. These three heuristic rules, in combination with a formal representation of product functions, provide a systematic approach to analyze the modularity of product families. Several product cases have been used to validate the method and construct a database. Following the same rationale, an analytical method was proposed to incorporate the customer demands into architecture building [Yu et al., 1999]. Moreover, a modular product architecture was developed to permit the platform to shift in size and type [Dahmus et al., 2000].

As an extension to the heuristic methods, a quantitative functional method was proposed for product architecture building [Stone et al., 2000a]. This method incorporates a functional model of the product and customer needs. Numerical manipulation was carried out to compute the customer need ratings with respect to function similarity. This leads to the possible combinations of functions to form modules.

These are representative approaches to establish product architectures based on the formal representation of function. They are helpful to speed up the process. However, the application of the quantitative methods and heuristic rules still relies heavily on human designers, which hinders their efficiency and consistency. Hölttä et al. [2003] proposed an algorithm to generate a modular product architecture based on a metric of module distance. This method uses quantitative measures and is supported by computational algorithms. Finally, computational tools, such as SA [Gu and Sosale, 1999] and GA [Yu et al., 2003], have been adopted to build modular product architecture.

In practice, modularity analysis based on experience and manual operations still abound. While this is effective for products which structures are well understood, it becomes cumbersome for complex products. Rapid and intelligent tools are required to facilitate modularity analysis and product architecture building. Hence, the product architecture still needs to be enhanced. First, the establishment of such an architecture should be consistent with the product information modeling schemes. Formal representation schemes should be used to build product architectures. Second, computational tools are required to enable rapid and intelligent building of product architectures. The next section introduces a neural network based approach to meet these requirements.

4.4 Product Architecture Building Using Self-Organizing Map

This section presents a computational method to build a FPA. The purpose of FPA is to identify the typical product functions of a product family. This process is called function analysis. As an important step in function analysis, the clustering of product functions is achieved using the self-organizing map (SOM) method.

4.4.1 *Introduction of SOM*

A SOM network is a special class of neural networks based on the theory of competitive learning [Kohonen, 1989]. Self organization refers to the evolution of a system into an organized status without external interference. The process is illustrated in Figure 4.6. A set of seemingly disordered input data is given in an arbitrary space (Figure 4.6(a)). The topology of the input data is generated by mapping it into a different space. Figure 4.6(b) shows the initial status of the system. After an ordering process, a new feature map is formed to reveal the intrinsic relations of the input data (Figure 4.6(c)). The essence of a SOM is its ability to identify the intrinsic statistical features contained in the input patterns and generate topographic maps (called feature maps) based on unsupervised learning. Unlike the supervised-learning neural networks, such as back-propagation (BP) and radial-basis function (RBF), a SOM network does not rely on the assignment of learning rules and training data. Instead, the SOM network is expected to identify the underlying rules of a given data set.

A SOM network structure usually consists of three layers, *viz.*, the input layer, the competitive layer and the output layer. Figure 4.7 illustrates a typical SOM model, *viz.*, the Kohonen model [Haykin, 1999]. The input layer accepts a multi-dimensional data pattern, which is usually represented as a vector. The competitive layer can be organized into 1- or 2-dimensions. Each neuron receives a summation of the weighted inputs from the input layer, and is associated with a collection of adjacent neurons, which form its 'neighborhood'. Once the network has been initialized, three procedures are involved in the formation of the feature map.

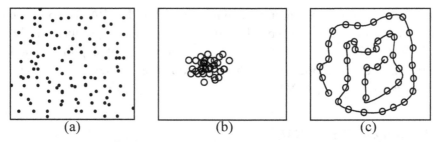

(a) Input data distribution, (b) Initial condition of the feature map, (c) Condition of the feature map after training.

Figure 4.6 System evolution as the formation of feature map

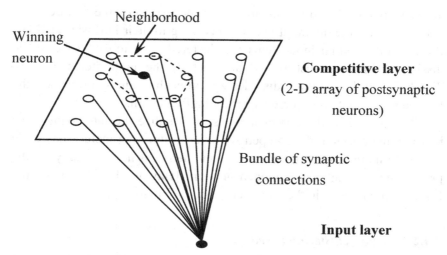

Figure 4.7 Self-organizing map: the Kohonen model* [Haykin, 1999]

(1) **Competition**: For each input vector, the neurons in the competitive layer compute their responsive values according to a distance function. The neuron with the largest responsive value is declared as the winner. In Figure 4.7, a black neuron is declared as the winner based on a pre-defined distance function, which is not necessarily the Euclidean distance to the input neuron. However, if the Euclidean distance is used, the winning neuron is usually selected as the one with the minimum distance.

(2) **Cooperation**: The topological neighbors of the winner are determined to provide the basis for cooperation among them. The winner and its neighbors are collectively called the excited neurons.

(3) **Synaptic activation**: The excited neurons increase their individual responsive values of the distance function in relation to the input vector. This is achieved through adjusting the weight vectors of the excited neurons such that they move towards the input vector.

At the initial stage of network formation, no specific order is present. However, after the training processes, the neurons in the competitive layer

---

* The output layer is not explicitly defined in this model.

are self-organized into a few meaningful patterns, *i.e.,* the feature maps which can arrange the input vectors according to their intrinsic relations. For example, similar input items are clustered close to each other while dissimilar ones are distributed far apart. If the input items are still distributed in the map in a disorder way after the training processes, the feature map is not considered as a meaningful pattern.

The purpose of the output layer is to visualize the interconnections between the nodes in the competitive layer. It does not include any logic for the formation of the feature maps, and hence, is not necessary for the proper functioning of a Kohonen network [Haykin, 1999]. The output layer is not included in the Kohonen model in Figure 4.7.

4.4.2 *Function clustering based on SOM*

SOM is used to perform unsupervised clustering of the product functions. It is expected that some patterns can be identified based on the similarity between the functions. The pre-requisite of function clustering is a set of products which function structures have been built according to the product information model. Such a function structure has been introduced in Chapter 3.

At the start of function clustering, all the atomic functions within the function structures belonging to different products are retrieved. The atomic functions are represented in the form of coded key element vectors (KEVs). The underlying principle is that the attributes contained in the key elements can be considered as coordinates that constitute a multi-dimensional discrete space. Hence, the atomic functions can be viewed as data points distributed in this space. If N attributes are used to represent the functions, an N-dimensional space can be constituted. This concept is illustrated with a 2D plane in Figure 4.8. The functions belonging to a set of related products are scattered in this space (the shaded circles). It is expected that similar functions from different products are topographically close to each other and intrinsically fall into a specific group (the ellipse). However, when many products are involved and the structural complexities of these products are diverse, the relationships between the functions are unclear. Thus, these groups are 'invisible' to a designer before any computational analysis.

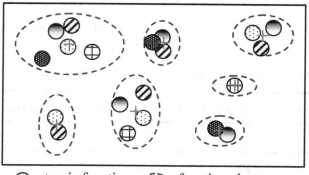

○ atomic functions ⟨ ⟩ function clusters
+ clustering center
Note: Same shade represents functions belonging to same product

Figure 4.8 Graphical interpretation of function clustering

A three-layer SOM architecture was built to carry out function clustering. The input data vector is derived from the KEV. Seven nodes are used to represent seven elements, *viz., function action, input energy flow, input material flow, input signal flow, output energy flow, output material flow*, and *output signal flow*. The KEV represents a function using 3-digit or 4-digit codes. However, the SOM neural networks require the input data to be within a range of [0, 1]. Hence, the input data must be normalized. Based on the coding schemes of the function actions and flows, the normalization is carried out in a straightforward way. A function action *xxx* is always in the range of [000, 999]. The normalized code is set as *xxx/1000*. Similarly, the normalized flow is set as *xxxx/10000*. Table 4.1 shows an example of the input vector for the function 'heat generation'.

Thus, an input vector can be represented as a vector $\vec{f}_{i[7\times1]}$, which elements are within the range of [0, 1]. Next, a 2D, *n*-by-*n* lattice is constructed in the competitive layer, where *n* is a positive integer depending on the scale of the problem. As a rule-of-thumb, n can be set as the average number of atomic functions of the products to be analyzed. Each node in the lattice is connected to the input nodes by a weight vector $\vec{w}_{j[7\times1]}$. There are a total of $l=n^2$ weight vectors in the SOM network.

Table 4.1 Input vector of an atomic function – 'heat generation'

Elements	A_F	$I_W(E_{NG})$	$I_W(M_{AT})$	$I_W(S_{IN})$	$O_W(E_{NG})$	$O_W(M_{AT})$	$O_W(S_{IN})$
Value	convert	electric wattage	—	electric signal	heat	—	—
Code	302	1202	0	3005	1304	0	0
Normalized code	0.302	0.1202	0	0.3005	0.1304	0	0

After all the input vectors have been imported, a training session is performed according to the three procedures discussed in Section 4.3.1. In the function clustering problem, the following steps are involved.

(1) Initialization: Random values are assigned to the initial weight vectors $\vec{w}_j(0)$.

(2) Matching: At training step p, the Euclidean distances between the input vector $\vec{f}_i$ and the weight vectors $\vec{w}_j(p)$ are computed. The winning neuron is selected as the one which weight vector has the minimum distance to the input vector.

$$\chi(\vec{f}_i) = \arg\min \left\| \vec{f}_i - w_j(p) \right\|, \quad j = 1, 2, \ldots, l \tag{4.1}$$

(3) Neighborhood activation: The neighbors of the winning neuron are selected according to their topographical distances to the winning neuron. For example, the winning neuron itself ($\chi(\vec{f}_i)$) is called

$N(0)$; the immediate neighbors of $\chi(\vec{f_i})$ are called $N(1)$, and so on (Figure 4.9). $N(i)$ are collectively called the excited neurons.

(4) **Updating**: The weight vectors $\vec{w}_j(p)$ of the excited neurons are updated using the following criterion.

$$\vec{w}_j(p+1) = \vec{w}_j(p) + \eta(p) h_{j,N(i)}(p) \left[\vec{f_i} - \vec{w}_j(p) \right] \qquad (4.2)$$

where $\eta(p)$ is the learning rate, and $h_{j,N(i)}(p)$ is the neighborhood function, which differs for the excited neurons that are located in different neighborhood $N(i)$. Thus, the weight vectors of the excited neurons can be moved slightly towards the input vector. Figure 4.10 illustrates the updating process in a 2D plane.

(5) **Continuation**: The training data is presented to the network repeatedly such that the synaptic weight vectors are updated continuously to resemble the input vectors.

Several trial training sessions can be carried out by varying the controlling parameters, such as the size of the lattice (n), the type of lattice (rectangle, hexagonal, random), training epochs, and the learning rates (η). A visual feature map is generated with the functions clustered at different nodes.

Finally, a designer is prompted to export the feature map into the output layer, which is organized as a two-level tree structure (Figure 4.11). A root node represents a cluster entry and the leaf nodes represent the actual functions assigned to this cluster. The tree structure will be further refined by human designers. The refinement process allows the empirical knowledge of the human designers to be incorporated. For example, possible errors that are caused either by insufficient training or by noise data can be identified. Insufficient training happens when the size of the lattice or the training epoch is too small. Noise data emerges when a designer fails to model a product function using suitable vocabularies. The refinement includes operations, such as merging similar functions, deleting non-representative functions, and assigning names to the clusters. The final outcome is a set of common functions for a product family, with related functions belonging to different product cases assigned to them.

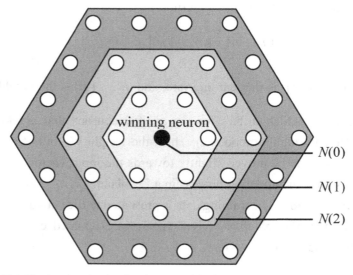

Figure 4.9 Neighborhood activation in a hexagonal lattice

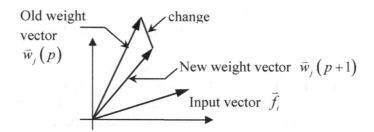

Figure 4.10 Updating weight vector in a 2D plane

The MATLAB® neural networks toolbox provides the basic tools to implement the algorithms of SOM. These tools are adopted in this book in the function analysis program.

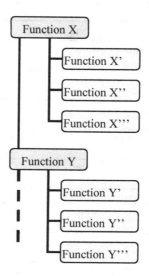

Figure 4.11 A tree-structure of the output layer

4.4.3 *A case study*

This section presents an example to build the FPA for a family of electric kettles using the SOM method. This example illustrates how SOM is used to cluster product functions and facilitate the building of product architecture. It should be noted that the structure of this product is very simple, and hence, it is not really difficult to build the FPA manually. However, for products which structures are more complicated, or when a large number of products are involved, the SOM method can be of significant advantage.

The information and data of four different electric kettles were collected. The function structure of each product was established through function decomposition (refer to Figure 2.12). The atomic functions of four different products are shown in Table 4.2. Accordingly, Table 4.3 lists the input vectors that have been retrieved from the normalized, coded KEV representation of the atomic functions. The coding scheme follows the taxonomies presented in Tables 3.4 and 3.5.

Table 4.2 Atomic functions of four sample products

P_i	NAME	A_F	I_w-E_{NG}	I_w-M_{AT}	I_w-S_{NG}	O_w-E_{NG}	O_w-M_{AT}	O_w-S_{IN}
	water fetching	convey	human force	water			water	
	heating control	turn-on			temperature			switch on
	show status	display			temperature			visual
1	heat generation	convert	wattage		switch on	heat		
	heat transfer	conduct	heat	water			water	
	keep warm	Store		water			water	
	contain water	enclose		water				
	water holding	enclose		water				
2	heating	convert	wattage	water	electric signal		water	
	heating control	turn-on			human operation			switch on
	heating control	turn-on			human operation			switch on
	generation heat	convert	wattage		electric signal	heat		
3	transfer heat	conduct	heat	solid-liquid			solid-liquid	
	warm keeping	Store		solid-liquid			solid-liquid	
	burn protection	constrain	heat			heat		
	heat control	turn-on			temperature			switch on
	produce heat	convert	wattage		electric signal	heat		
4	transfer heat	convey	heat	water			water	
	displace status	display			temperature			visual
	keep warm	Store		water			water	
	fetch water	channel	human force	water			water	

Table 4.3 Normalized input data of the atomic functions

P_i	NAME	A_F	I_w-E_{NG}	I_w-M_{AT}	I_w-S_{NG}	O_w-E_{NG}	O_w-M_{AT}	O_w-S_{IN}
	water fetching	0.404	0.1108	0.2102	0	0	0.2102	0
1	heating control	0.503	0	0	0.3102	0	0	0.3006
	show status	0.506	0	0	0.3102	0	0	0.3106
	heat generation	0.302	0.1202	0	0.3006	0.1304	0	0
	heat transfer	0.403	0.1304	0.2102	0	0	0.2102	0
	keep warm	0.143	0	0.2102	0	0	0.2102	0
	contain water	0.905	0	0.2102	0	0	0	0
2	water holding	0.905	0	0.2102	0	0	0	0
	heating	0.302	0.1202	0.2102	0.3005	0	0.2102	0
	heating control	0.503	0	0	0.3	0	0	0.3006
	heating control	0.503	0	0	0.3	0	0	0.3006
3	generation heat	0.302	0.1202	0	0.3005	0.1304	0	0
	transfer heat	0.403	0.1304	0.2301	0	0	0.2301	0
	warm keeping	0.143	0	0.2301	0	0	0.2301	0
	burn protection	0.902	0.1304	0	0	0.1304	0	0
	heat control	0.503	0	0	0.3102	0	0	0.3006
	produce heat	0.302	0.1202	0	0.3005	0.1304	0	0
4	transfer heat	0.404	0.1304	0.2102	0	0	0.2102	0
	displace status	0.506	0	0	0.3102	0	0	0.3106
	keep warm	0.143	0	0.2102	0	0	0.2102	0
	fetch water	0.402	0.1108	0.2102	0	0	0.2102	0

Seven nodes were used in the input layer and a 5-by-5 hexagonal lattice was used in the competitive layer. In the lattice, each dot represents a grid node (Figure 4.12), and this node is connected to the input nodes by a weight vector $\bar{w}_{j[7\times 1]}$. Thus, there are 25 weight vectors altogether. The position of a node is denoted as $N(i, j)$, where i and j are integers, and $1 \leq i, j \leq 5$. For example, the origin is denoted as $N(1, 1)$. Note that this is different from the Cartesian coordinates.

When the input data (the atomic functions) were initially imported, they were distributed in the lattice according to the initial values of the weight vectors (Figure 4.12). The map does not reveal any order in the input vectors. Next, the training process was carried out according to the steps discussed in Section 4.2.2. Experiments were carried out repeatedly with different settings of the learning rates and training epochs. It was found that similar feature maps were built. A typical pattern after the training process is shown in Figure 4.13.

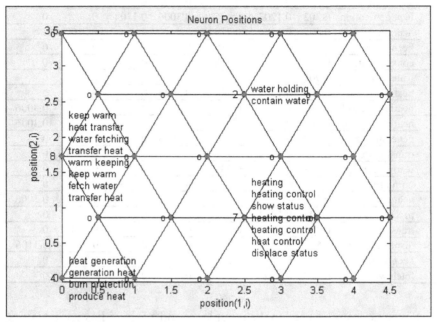

Figure 4.12 Initial status of the competitive layer

As can be seen from the feature map, the functions have been grouped according to their affinity to each other. The number on the left of each node represents the total number of functions that have been clustered at that node. For example, three functions are located at $N(1, 3)$, and they are reasonably considered as a group denoting 'heat generation'. Based on this pattern, a tree structure was automatically generated as the output layer, where a node of this tree must have at least one function assigned (Figure 4.14(a)). This tree structure was further refined by human designers. For example, the 'heating' function located at $N(1, 1)$ is similar to 'heat generation'. As a result, it was merged to the 'heat generation' cluster located at $N(1, 3)$. Finally, seven clusters were identified for the electric kettle products (Figure 4.14(b)). The clusters represent the common functions for a product family, which are denoted as:

$$f = \begin{bmatrix} f_1 \\ f_2 \\ f_3 \\ f_4 \\ f_5 \\ f_6 \\ f_7 \end{bmatrix} = \begin{bmatrix} \text{keep warm} \\ \text{heat transfer} \\ \text{water fetching} \\ \text{heat generation} \\ \text{disply status} \\ \text{water holding} \\ \text{heating control} \end{bmatrix}.$$

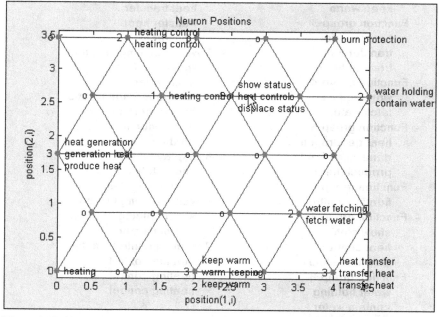

Figure 4.13 Clustering pattern in the competitive layer after training

4.4.4 *Evaluation of the SOM method*

The formation of the feature map requires the function base of the product information model. Function similarity is the basis to cluster the functions, where function similarity is estimated based on the coded functions and flows. These are analogous to the quantitative and heuristic methods used to build modular architectures [McAdams *et al.*, 1999; Zamirowski and Otto, 1999; Stone *et al.*, 2000a, 2000b]. The resulting function clusters are similar to the product platforms obtained in these methods. However, the SOM method adopts a process that is different from these methods. In particular, unsupervised learning is used for function analysis.

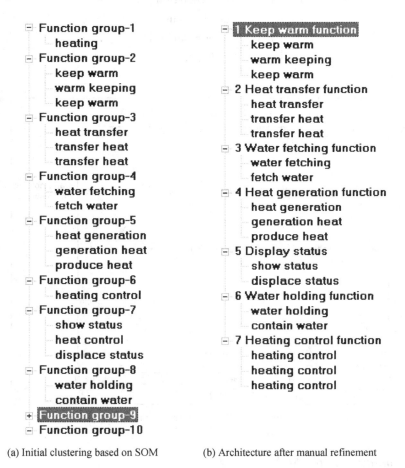

(a) Initial clustering based on SOM   (b) Architecture after manual refinement

Figure 4.14 FPA of the electric kettle products

SOM is a necessary step although a refinement process can be carried out subsequently by a human designer. The reasons are, firstly, the SOM method analyzes the data based on the KEV while a designer relies more on empirical information; secondly, a manual analysis of the data from the KEV will involve significant effort, especially when a large number of functions are included; and thirdly, the designer may not be aware of the appropriate numbers or patterns of the clusters at the beginning of the function analysis. Therefore, the advantages of using the SOM method are: (1) an expedition of the process, (2) an alleviation of human labor, (3) determination of useful initial knowledge and patterns of the architecture, and (4) an analysis of the data from a perspective other than empirical observation. Therefore, the SOM process and human refinement are complementary to each other.

To summarize, the SOM method can generate a feature map from a set of functions without human supervision. Thus, it can identify the preliminary patterns of the functions and facilitate the building of product architecture. This method is different from product platform design using the top-down approaches which require tremendous product analysis. In comparison with modular design approaches, such as the QFD-based approach, DSM-based approach, and the heuristic/quantitative approach, the SOM method is a computational technique using unsupervised learning algorithms. Therefore, SOM has less reliance on human expertise. Although the final formation of the product architecture still requires refinement by human designers, the refinement process is carried out after a feature map has been formed such that a rudimentary architecture is already present. Therefore, the refinement process will not pose a heavy load on the designer. Table 4.4 summarizes the performance of SOM in comparison with the legacy methods.

Table 4.4 Comparison of methods to product architecture building

Method	Computation tool	Computation rigor	Repeatability	Complexity of products	Human labor
QFD-based	HOQ and variants	Low	Poor	Low	High
DSM-based	DSM	Medium	Good	High	Medium
Heuristic/ quantitative	Function structure; database	Medium	Fair	Medium	Medium
SOM	ANN (SOM)	High	Good	Medium	Low

4.5 Other Relevant Issues in Product Platform Design

As the core of product platform, a product architecture deals with the mapping relationship between FRs and DPs. However, to effectively carry out product family design, a product platform must also include various types of knowledge and the association between knowledge pieces to provide decision support. This section presents the other relevant issues in product platform design.

Typically, design involves the process of finding the proper DPs to fulfill the design requirements. Therefore, it is important to: (1) identify the DPs, (2) identify relevant design requirements with suitable measurements, and thereafter, (3) establish the relationships between the design requirements and the DPs. To achieve these, a few knowledge extraction operators, denoted as $\{Op_x\}$, are developed. The overall process is organized as a domain mapping process similar to the zig-zag decomposition in axiomatic design [Suh, 2001] (Figure 4.15). Among these operators, Op_f is the function analysis method to establish the FPA and has been presented in Section 3.3. Op_k and Op_r are discussed in this section, and Op_i and Op_c will be discussed in the subsequent chapters.

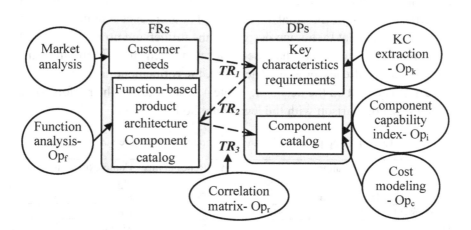

Figure 4.15 Mapping route form design requirements to design parameters

Design of Product Platform

The top-level design requirements are the customer needs (CNs), which include the most important product features that define the market segmentations. Market analysis can be carried out to obtain a set of CNs. Since this book focuses on the engineering aspects, it is assumed that the CNs are known, and are denoted as $r = [r_1, r_2, ..., r_s]^T$. For example, the customer requirements of an electric kettle are:

$$r = \begin{bmatrix} r_1 \\ r_2 \\ r_3 \\ r_4 \\ r_5 \end{bmatrix} = \begin{bmatrix} \text{energy usage} \\ \text{capacity} \\ \text{safety} \\ \text{ease of use} \\ \text{cost} \end{bmatrix}.$$

The CNs are quantified by KCs. The extraction of a set of KCs to signify the performance features and define a product family is discussed in Section 3.4.1. The KCs are dependent on a set of functions, which is defined in the FPA. The bottom-level DPs are the physical components, which are contained in the component catalog. The formation of the component catalog is discussed in Section 3.4.2. From the design reuse perspective, a practical information processing strategy is illustrated in Figure 4.16, which incorporates (1) the extraction of KCs, (2) the establishment of FPA, and (3) the formation of component catalog.

4.5.1 Extraction of KCs as performance criteria

As mentioned earlier, each product (p_i) contains a set of KCs: $K_i^0 = [k_{i1}, k_{i2}, ..., k_{ip}]^T$. The set of KCs may be different across different products. However, since the products collected in the same general design space (SP$_i$) are similar to each other and are expected to form a product family, it is possible to identify a set of common KCs to measure the performance of the products. This process is carried out manually. The resulting set of common KCs is denoted as $k = \begin{bmatrix} k_1, k_2, ..., k_p \end{bmatrix}^T$. As an example, the KCs for the electric kettle product family include:

$$k = \begin{bmatrix} k_1 \\ k_2 \\ k_3 \\ k_4 \\ k_5 \\ k_6 \\ k_7 \end{bmatrix} = \begin{bmatrix} \text{power consumption} \\ \text{dimension} \\ \text{water capacity} \\ \text{automatic control} \\ \text{cost} \\ \text{MTBF} \\ \text{water fetching method} \end{bmatrix}.$$

In the subsequent design synthesis and evaluation stage, k is used as the major performance criterion that differentiates the product members in a product family.

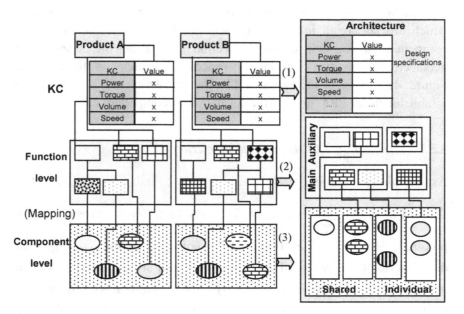

Figure 4.16 Establishment of product platform based on design reuse

4.5.2 Formation of component catalog

Each product (p_i) contains a set of physical components: $M_i^0 = [m_{i1}, m_{i2}, ..., m_{in}]^T$. Components that belong to different products can be collected into the component catalog and will be used as the DPs in the design synthesis and evaluation stage. The component catalog is organized as a set of slots that correspond to the common functions defined by the FPA. Components that implement the same product function are assigned to the same slot. Thus, the relationships between the components and the functions are retained. The components are collectively denoted as:

$$m = [s_1, s_2, ..., s_m]^T$$

where s_i is the ith slot which corresponds to the common function f_i, and

$$s_i = [m_i^1, m_i^2, ..., m_i^{n_i}]^T$$

where n_i is the number of components in this component slot.

Other than the attributes that characterize each component, the component cost and the performance capability should also be identified. Cost can be identified based on historical data. A cost road-map is established for each component. Performance capability refers to the capability of a component to satisfy specific design requirements. It is extracted using the component capability index operator (Op_i), as will be discussed in Chapter 7.

4.5.3 Establishment of mapping route using correlation matrices

Based on the above discussions, the DPs and design requirements at different levels can be defined. Given that $r_{s \times 1}$, $k_{p \times 1}$, $f_{m \times 1}$, $m_{n \times 1}$ represent the vectors of CNs, KCs, common functions, and physical components in the component catalog, respectively, a mapping route can be established using a few correlation matrices, which constitute the Op_r process.

$$r = TR_1 \times k \qquad (4.3)$$

$$k = TR_2 \times f \qquad (4.4)$$

$$f = TR_3 \times m \qquad (4.5)$$

TR_1 is an $s \times p$ matrix which element (i, j) is either '1' indicating r_i is related to k_j, or '0' indicating r_i is not related to k_j. A similar definition is made for TR_2 and TR_3. TR_1 and TR_2 can be designed manually, similar to the QFD processes. In addition, TR_2 can be assisted by the X_{K-F} relationship defined in the contextual facet of the product information model. TR_3 is a natural outcome of the establishment of the component catalog. Using the electric kettle as an example, r, k, and f, have been defined earlier. TR_1 and TR_2 are established as shown in Tables 4.5 and 4.6, respectively. The overall mapping route is illustrated in Figure 4.17.

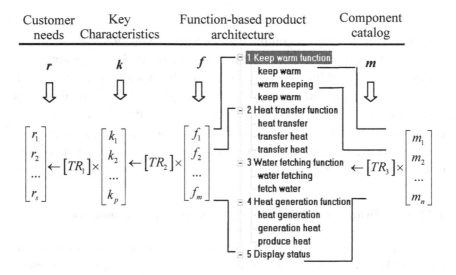

Figure 4.17 Mapping route from CNs to physical components

Table 4.5 Correlation between customer needs and KCs (TR_1)

TR_1		k_1	k_2	k_3	k_4	k_5	k_6	k_7
r_1	Energy usage	1	0	0	0	1	0	0
r_2	Capacity	0	1	1	0	1	0	0
r_3	Safety	0	0	0	1	1	1	0
r_4	Ease of use	0	0	0	1	0	0	1
r_5	cost	0	0	0	0	1	0	0

k_1: power consumption k_2: dimension k_3: water capacity
k_4: automatic control k_5: cost k_6: MTBF
k_7: water fetching method

'1' indicates that a CN is dependent on a KC; '0' otherwise.

Table 4.6 Correlation between KCs and functions (TR_2)

KCs		f_1	f_2	f_3	f_4	f_5	f_6	f_7
k_1	power consumption	0	1	0	1	0	0	0
k_2	dimension	0	0	0	0	0	1	0
k_3	water capacity	0	0	0	0	0	1	0
k_4	automatic control	1	0	0	0	0	0	1
k_5	cost	1	1	1	1	1	1	1
k_6	MTBF	1	1	1	1	1	1	1
k_7	water fetching	0	0	1	0	0	0	0

f_1: Keep warm f_2: Heat transfer f_3: Water fetching
f_4: Heat generation f_5: Display status f_6: Water holding
f_7: Heating control

4.6 Summary

Product platform defines the technology core that connects the functional domain with the physical domain. From the top-down and bottom-up perspectives, two streams of platform-building approaches have been prevalent, *viz.*, the scale-based approach and the module-based approach. Existing approaches in this area relies heavily on human intelligence, which makes the process slow and unrepeatable. Computational tools are required to enable rapid and intelligent building of product platform.

The construction of product architecture is a central issue in product platform design. The SOM method is proposed as a novel tool to build the modular product architectures. The SOM method is an unsupervised learning technique in neural networks. It is especially useful to identify the latent design patterns underneath the seemingly chaotic input data. Using the SOM method, the functions of different products can be clustered according to their similarity. Thus, a modular product architecture can be easily established. This process can be carried out using computational algorithms, which is fast and less reliant on human effort.

Based on the FPA, a product platform can be built in the form of a mapping route from the CNs to the physical components. The product platform can be used to derive product variants using automated design synthesis methods. This is the topic to be addressed in the next chapter.

Chapter 5

Optimization in Product Design

Previous chapters have discussed the techniques to model and analyze product information such that it can be reused in new designs. This chapter will discuss the design-by-reuse issue. In particular, the design activities to create new products are the major concern. These design activities are called design synthesis. Typical in a design synthesis problem, the solution configuration is generated and evaluated with respect to one or a few design objectives and constraints. Such a problem can be solved using various optimization algorithms. This chapter focuses on the optimization in product configuration design. The fundamental concepts of engineering optimization are presented. A few widely adopted optimization algorithms are studied in the context of product configuration design. A genetic algorithm (GA)-based method, namely multi-objective struggle genetic algorithm (MOSGA), is elaborated with a case study to show its power in solving the configuration design with multi-objectives.

5.1 Introduction

In engineering design, the quality of a solution cannot be easily captured in a single criterion. For example, picture quality and product cost are two basic criteria to determine the characteristics of a TV set. A designer may want to maximize the picture quality and minimize the product cost at the same time. However, more often than not, such objectives are conflicting to each other, *i.e.*, the improvement in one objective inevitably causes the deterioration of the other. This is a typical situation faced in engineering design. An optimization problem based on multiple objectives is called a multi-objective optimization problem (MOOP). The techniques to solve a MOOP must deal with optimal trade-offs between the conflicting objectives.

A general formulation of MOOP is given in the following equations.

$$\text{Min } T(x) = [t_1(x), t_2(x), \ldots, t_r(x)] \tag{5.1}$$

$$\text{s.t. } x = [x_1, x_2, \ldots, x_n]^T, \ x \in S \text{ and} \tag{5.2}$$

$$g_j(x) \leq 0, \ j = 1, 2, \ldots, l \tag{5.3}$$

where $t_1(x), t_2(x), \ldots, t_r(x)$ are the r objective functions. The values of $T(x)$ constitute an attribute/criteria space, i.e., $T(x) \in R^r$. $x = [x_1, x_2, \ldots, x_n]^T$ is the vector containing n DPs. S is the parameter/solution space, and $S \in R^n$. Usually, S can be divided into two regions, namely the feasibly space, denoted as $|S|$, and the infeasible space, according to a set of constraints $g_j(x) \leq 0$ (for $j=1,2,\ldots,l$). It should be noted that Equation (5.3) is a generalized representation of the constraints, such that when '≥' or '=' are involved, the equation needs to be adjusted. Figure 5.1 shows the mapping from the parameter space to the attribute space, where two variables and two objective functions are involved.

Let t_1^*, t_2^*, …, t_r^* be the individual minima of each objective function, respectively. A utopian ideal solution is defined as $T^* = [t_1^*, t_2^*, \ldots, t_r^*]$. However, such an ideal solution is rarely attainable in reality. Therefore, a useful replacement is the Pareto-optimal solution which is also called a non-dominated solution or non-inferior solution. A solution is said to dominate another if it is superior in most, if not all the objectives. Considering the minimization problem $T(x)$ with r objectives, and two solutions x_1 and x_2, x_1 are said to dominate x_2 if

$$\forall i \in \{1,2,...,r\} : t_i(x_1) \le t_i(x_2) \text{ and } \exists j \in \{1,2,...,r\} : t_j(x_1) < t_j(x_2) \quad (5.4)$$

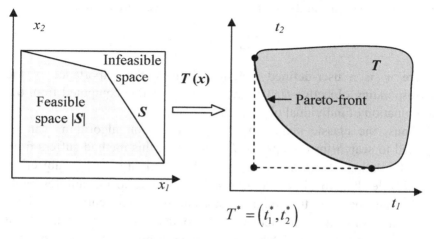

Figure 5.1 The parameter space, attribute space and Pareto-front

Thus, a solution is said to be a Pareto-optimal if it is not dominated by any solution in the feasible space $|S|$. Typically, the Pareto-optimal solutions constitute a curve or surface, called the Pareto-front, which shape signifies the nature of the trade-offs between different objectives (Figure 5.1).

To solve a MOOP, a number of algorithms have been reported. Some of the algorithms have converted a MOOP into a single-objective optimization problem (SOOP) using a few user-defined procedures. While these methods have the advantage of producing a single trade-off solution, bypassing the need for further interactive decision-making, it may conceal valuable alternatives during the search process, and hence, possibly necessitate extra search efforts, and lead to sub-optimal solutions [Fenseca and Fleming, 1998]. Another stream of algorithms treats the objectives separately. Accordingly, the entire set of Pareto-optimal solutions is identified. Final solutions are selected from the Pareto-optimal solutions based on meta-knowledge of the designed system.

A few widely adopted methods for solving MOOP are discussed next.

5.1.1 *Weighted sum method*

Equation (5.5) is a standard formulation of the weighted sum method.

$$T_w(x) = \sum_{i=1}^{r} w_i t_i(x), \quad w_i > 0 \tag{5.5}$$

where w_i is a user-defined weight indicating the importance of the corresponding objective $t_i(x)$. $T_w(x)$ is a scalar value computed from the combination of individual objective functions.

Thus, the classic single-objective optimization algorithms can be applied to search for the optimal T_w. However, this method suffers from three limitations. First, the assignment of the weight vector is subjective, possibly leading to biased solutions. Second, due to the unforeseeable relationship between the weight vector and the Pareto curve, a uniform distribution of weight parameters rarely produces a uniform distribution of points on the Pareto set [Deb, 2001]. Often, a majority of solutions are clustered in certain regions of the Pareto set with few solutions in the interesting middle region. As such, the solution provides little insight into the shape of the Pareto-front curve. Third, this method is not efficient in solving non-convex problems.

There are a few variants of the weighted sum method, such as the weighted metric method and the Benson's method [Deb, 2001]. Instead of using the weighted summation of the objectives, these methods use other aggregation methods to combine the objectives into a single objective. In essence, they are similar to the weighted sum method.

5.1.2 *Goal programming*

The gist of goal programming (GP) is to find solutions that reach a pre-defined target for one or a few objective functions. If the targets cannot be strictly attained, the task is reduced to finding solutions that are closest to the targets. In GP, a distinction is made between an 'objective' and a 'goal'. An objective is a function to be optimized via changes in the problem variables; while a goal is a function with a target value or aspiration level, upon the fulfillment or adjacency of which, a solution is considered acceptable [Ignizio, 1982]. Accordingly, a general MOOP can be re-formulated in such a way that only one objective is to be minimized, and the remaining objectives are constrained to be less than the given target values. A classic GP model is given in Equation (5.6) [Charnes and Cooper, 1977].

$$\text{Minimize: } Z = \sum_{j=1}^{m}\left(d_j^+ + d_j^-\right) \quad (5.6)$$

$$\text{s.t. } \sum_{i=1}^{n} a_{ij}x_i - d_j^+ + d_j^- = b_j, \text{ for } j = 1,\ldots,m$$

$$d_j^+, d_j^-, x_i \geq 0, \text{ for } i = 1,\ldots,n; \ j = 1,\ldots,m$$

where d_j^+ and d_j^- are the positive and negative goal deviations, respectively. A goal deviation refers to the difference between the achieved objective and the desired objective. a_{ij} (where $i=1,\ldots,n$; $j=1,\ldots,m$) is the technological coefficients that represent the impact the decision variables x_i has on the right-hand-side coefficient, namely, the goals b_j. Thus, the GP model converts the optimization problem into the task of finding the solution that is 'least deviated' from the goals.

The GP model is very concise and a GP problem can be solved using well-established theories, such as linear and nonlinear programming. A large number of GP applications have been reported in diverse fields [Schniederjans, 1995]. However, it is not always easy to choose appropriate goals for the constraints. Moreover, GP has been criticized for being not Pareto efficient, *i.e.*, it cannot be used to generate the Pareto set effectively, particularly if the number of objectives is greater than two.

5.1.3 *Multi-level programming/rank ordering*

Multi-level programming aims at finding one 'optimal' point instead of the entire Pareto-front. Firstly, the objective functions are rank-ordered in terms of their relative importance. Next, a set of points $x \in C_1$ is found for which the minimum value of the first objective function is attained. Similarly, a set of points $x' \in C_2$, where $C_2 \subseteq C_1$, is found, which minimizes the second most important objective. The method proceeds recursively until all the objectives have been optimized on successively smaller sets.

Multi-level programming is effective if the hierarchical order among the objectives is of prime importance and the continuous trade-off among the functions is not a major concern. However, problems in the lower level of the hierarchy become very tightly constrained and often become infeasible, so that the objectives with low orders have minimal influence on the final solution. Hence, multi-level programming should not be used to find a sensible compromise solution among the various objectives which differences are subtle.

5.1.4 *Genetic algorithms*

The basic principles of GA have been introduced in Chapter 2. This section further discusses the GA-based techniques for solving MOOP. These approaches are usually divided into the non-Pareto and Pareto-based approaches [Fonseca and Fleming, 1998]. The characteristics of these two types of approaches are discussed next.

5.1.4.1 Non-Pareto-based approach

Vector Evaluating Genetic Algorithm (VEGA) [Schaffer, 1985] is a non-Pareto-based approach and is the first multi-objective GA. In VEGA, each objective is used in turn as the metric for selecting a sub-population. A new generation of non-dominated individuals is produced using crossover and mutation based on the sub-population. Fourman [1985] proposed a GA to search for multiple non-dominated solutions concurrently. However, a restriction of the non-dominated approaches is that the Pareto-optimal set tends to be ill-posed on the Pareto-front, *i.e.*, some parts of the Pareto-front are much more 'populated' than the other parts [Andersson and Wallace, 2002].

5.1.4.2 Pareto-based approach

The Pareto-based approach makes use of the concept of 'rank' according to Pareto-optimality [Goldberg, 1989]. In this approach, the fitness of an individual is represented as its rank, and it is derived from non-dominated sorting. The procedures to assign ranks based on non-dominated sorting are shown in Figure 5.2. Srinivas and Deb [1995] developed the non-dominated sorting GA (NSGA) based on this approach. Another way to rank the population is to compute the degree of dominance [Fonseca and Fleming, 1998]. The rank of an individual is the number of population members that dominate over it plus one. Therefore, a lower rank indicates a better fitness, and the solutions with the rank '1' are the non-dominated solutions. The assignment of ranks based on the degree of dominance is illustrated in Figure 5.3.

Andersson and Wallace [2002] proposed a Pareto-based approach, namely the MOSGA, and compared it with a few typical multi-objective GA, such as VEGA, NSGA, multi-objective GA (MOGA), etc. MOSGA combines the struggle crowding GA [Grueninger and Wallace, 1996] and Pareto-based ranking [Fonseca and Fleming, 1998]. It is claimed that the MOSGA method can handle multi-modal attribute spaces with improved robustness. Moreover, it requires less tuning of the GA parameters in comparison with the other methods. These are desirable for the design synthesis problem. Therefore, the MOSGA method for configuration

120 *Design Reuse in Product Development Modeling, Analysis and Optimization*

design synthesis is elaborated in this book. Before the discussion of MOSGA in design synthesis, it is worthwhile to define the design synthesis problem, namely the product configuration design problem.

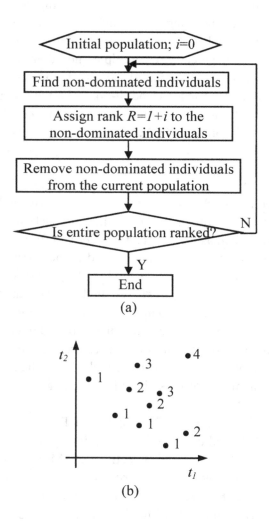

Figure 5.2 Population rank based on non-dominated sorting

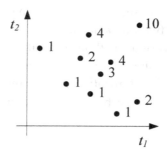

Figure 5.3 Population rank based on the degree of dominance

5.2 Automated Design Synthesis

From the reuse perspective, product design is a process of synthesizing product configurations from existing components based on the product platform. This includes direct retrieval of relevant modules and synthesis of products by combining a set of modules. The process can be carried out manually or automatically. To achieve efficient product design, automated design synthesis is required. Automated design synthesis is especially useful for solving large combinatorial problems, such as configuration design. Design synthesis can be carried out using various computational tools, such as agent-based methods, GA, simulated annealing (SA), branch-and-bound (B&B) method, etc.

5.2.1 *Configuration design*

Configuration design features a broad variety of design problems. A general configuration design task is characterized by (1) the generation of solutions, (2) satisfying a set of design requirements, based on (3) "a fixed, predefined set of components, where a component is described by a set of properties and ports for connecting it to other components" [Mittal and Frayman, 1989]. This definition has been widely adopted and extensively investigated in literature [Yu and MacCallum, 1995; Wielinga and Schreiber, 1997; Corbett and Rosen, 2004]. In engineering design, a proper formulation of the design problem and an effective exploration of the design space are the two fundamental issues in configuration design.

In the next section, various methods to deal with design synthesis for product configuration design are studied. These methods involve the problem formulation and optimization algorithms to search the design space. Various algorithms have been used to increase the efficiency of design synthesis.

5.2.2 *Design synthesis techniques*

5.2.2.1 Exhaustive search method

Exhaustive search represents an effort to visit and evaluate each and every solution in the design space. Hernandez *et al.* [2003] formulated the product family design problem as a problem of access in a geometric space, where exhaustive search was used to access the geometric spaces. Ong *et al.* [2006] formulated product customization as a Constraint Satisfaction Problem (CSP), which was solved using the invasion-based algorithm, basically an exhaustive search.

It is obvious that exhaustive search has limited power and is applied only when the design space is very small. In cases where the design space is large, more effective search algorithms are required.

5.2.2.2 Graph-based method

Graph-based method is, in a strict sense, not an optimization method. It is discussed here because (1) it represents a unique way to formulate the design synthesis problem, and (2) design synthesis is carried out based on specific graph grammars. In essence, graph-based methods are based on formal, graph-based representations of components, interfaces, and systems. The validity and integrity of a solution is largely dependent on the definition of the graph grammar.

Bond graph is a widely adopted generic method for modeling electrical, magnetic, mechanical, hydraulic, pneumatic and thermal systems [Vjekoslav and Montgomery, 2003]. It is developed based on the flow of energy or information within a system, and can be used to model systems from different domains consistently. Based on a set of simple elements (*e.g.*, effort/flow source, dissipater, capacitor, inertia, '0' and '1' junctions, gyrator and transformer), bond graph allows models to be translated into differential equations or computer simulation schemes. Hence, design synthesis can be carried out in an elegant manner.

Ulrich and Seering [1989] used bond graph chunks to specify a function. Thus, design synthesis can be carried out using a vector specification of the input and output functions. Bond graph provides a mechanism to generate designs from a specification of the desired behavior. Computer programs were written to implement the design synthesis process. However, it did not fulfill the requirement of automated transformation and subsequent modification of the designs. Malmqvist [1994] expanded the bond graph chunks approach to a function-based synthesis method, where a black box model is capable of creating a set of alternative designs that contain components to fulfill the desired functional and spatial characteristics.

Du *et al.* [2002a] proposed the Programmed Attribute Graph Grammar (PAGG) to specify the design space and assist product family generation. Furthermore, a graph rewriting program was developed that enables the derivation of product variants through graph transformations [Du *et al.*, 2002b]. However, the graph transformation operators for generating product family are not adequate to deal with complex product family configuration design with good efficiency. Such an inadequacy is actually a major limitation of the graph-based approaches because the generation of product configurations based on a limited number of graph transformation operators falls short of being too restrictive. These approaches require further development to deal with more complex product family design problems.

5.2.2.3 Linear and non-linear programming

Linear programming (LP) problems are optimization problems in which the objective functions and the constraints have linear relationships with a set of design variables. For example, given a vector of design variables

$x = [x_1, x_2, \ldots, x_n]^T$, and an objective function $f(x) = c_1 x_1 + c_2 x_2 + \ldots + c_n x_n = c^T x$, a LP problem can be represented as:

$$\text{Min: } c^T x \tag{5.7}$$
$$\text{s.t.: } Ax \leq b, \ x \geq 0$$

where c is a n-vector of objective coefficients, A is a $m \times n$ matrix, and b is a m-vector of constraint coefficients.

Similarly, nonlinear programming (NLP) is the process of solving an optimization problem with equalities and inequalities in which the objective function and constraints have nonlinear relationships with the design variables.

Accordingly, to solve the product configuration design problem using LP or NLP, first, the design problem must be formulated such that the design objectives can be formulated as linear/nonlinear functions of the design variables (the attributes of design components). Next, the classical LP and NLP algorithms can be used to search the design space. Many analytical and computational algorithms have been developed to solve the LP and NLP problems, such as Simplex and its variants for solving single objective optimization, and Multiplex for multi-objective optimization [Ignizio and Cavalier, 1994].

The LP and NLP methods have been used in a number of product configuration design problems, as listed in Table 5.1. Usually, the problem is defined in the continuous design space, and accordingly the problem can be formulated as sequential linear programming (SLP), sequential quadratic programming (SQP), generalized reduced gradient (GRG), etc. [Avriel, 1976]. Moreover, for quicker convergence, many LP and NLP are based on a gradient-based search, *i.e.*, they involve the computation of the derivatives of the objective functions to determine the search direction. In such situations, the search space must be continuous and differentiable.

However, configuration design is combinatorial in nature, and is characterized by a mixed discrete-continuous design space [Simpson *et al.*, 2001]. Usually, the design problem cannot be effectively represented in polynomial forms [Ignizio and Cavalier, 1994]. Hence, to use LP or NLP in configuration design, significant effort is required to properly formulate the design problem as a mix-integer programming problem, which is more difficult to solve. To tackle this problem, heuristic programming, such as GA and SA are promising.

5.2.2.4 Generic algorithm

GA has been advocated by many researchers, due to its power in solving complex combinatorial problems, for the design synthesis problem (Table 5.1). For example, Li and Azarm [2002] used GA to maximize designer's utility in the design evaluation and selection stage of product family design. D'Souza and Simpson [2003] proposed a GA method for solving product family design and optimization. In essence, for the configuration design problem, the following issues must be addressed to ensure the efficiency of GA.

(1) Formulation of design problem, including the representation of the product configuration structure, encoding of the compositional elements (design variables), and modeling of the design constraints;
(2) Choosing proper performance or cost factors as the objective function(s);
(3) Establishment of the relationship between the objective function(s) and the design variables;
(4) Strategy of controlling the search process, including population size, termination conditions, crossover/mutation rate, etc.

In comparison with other methods such as LP and NLP, GA is derivative-free. Therefore, it can deal with the discrete design space with ease. Moreover, a group of solutions (population) is maintained during the search process. Hence, it is possible to obtain multiple solutions. This is especially useful for designers who want to obtain a few candidate solutions, with each excelling in certain aspects.

On the other hand, in GA, the scale of the problem can be prohibitively large when many design variables are involved [D'Souza and Simpson, 2003]. The computation time increase proportionately with the number of design variables. Moreover, the diversity of the population is an important factor to ensure global or near global optimality. However, to maintain the population diversity is a tricky task. It may require considerable expertise or prolonged trial-and-error process. Finally, GA has been criticized for its inadequacy to refine a local search [Ishibuchi *et al.*, 1994; Yen *et al.*, 1998].

5.2.2.5 Simulated annealing

SA is a programming method that attempts to simulate the physical process of annealing in metallurgy. Unlike conventional optimization methods, such as hill-climbing, the search process in SA is characterized by randomly generated new solutions (neighbors) and probability-based acceptance of the new solution. A control parameter called 'temperature' is used to manage the probability. In particular, the temperature decreases gradually such that the probability of accepting an inferior solution is reduced as the search proceeds. Therefore, at the initial stage of SA, the search region is to a certain extent global. As the search proceeds, the search region is gradually limited to a local region. Hence, SA is suitable for dealing with multi-peak optimization problems, and can arrive at global or near global optimal solutions.

SA has been used in a number of product configuration design problems. For example, Fujita *et al.* [1999] proposed a modular design approach for product family configuration design, where SA was used to search for the optimal solutions. The design variables are represented as 0-1 integers and the optimization objective is the production cost. Hernandez *et al.* [2001] formulated the product family design problem as a compromise decision support problem (DSP), which is solved using SA based on the commercial optimization software OptDesX. Table 5.1 lists more applications of SA in product configuration design. Basically, the algorithm performs well in finding global optimal solutions. However, SA suffers from a few deficiencies:

(1) The setting of the control parameters, such as the initial temperature and the cooling rate is problem-specific and requires expertise.
(2) A high initial temperature and a slow cooling rate are needed to arrive at global optimal solutions. However, this may significantly reduce the computational efficiency of SA.

(3) The search process of SA is sequential rather than parallel. Only one solution is maintained during the process. This presents a barrier to applying parallel computing [Wang et al., 2005].

Table 5.1 Optimization algorithms for configuration design synthesis

Approach	Exhaustive	Graph	SLP	SQP	B&B	GRG	NLP	GA	SA	Agent
Campbell et al. [1999]								x	x	x
Chidambaram and Agogino [1999]							x			
Dai and Scott [2004]				x						
Darr et al. [1994]										x
de Weck et al. [2003]				x						
Du et al. [2002a]		x								
Du et al. [2002b]		x								
D'Souza and Simpson [2003]							x			
Fellini et al. [2000]						x				
Fujita et al. [1998a]				x						
Fujita et al. [1999]								x		
Fujita and Yoshida [2001]				x	x			x		
Fujita and Yoshioka [2003]								x		
Gonzalez-Zugasti et al. [2000]							x			
Gonzalez-Zugasti & Otto [2000]								x		
Hernandez et al. [2001]									x	
Hernandez et al. [2003]	x									
Jiang and Allada [2001]				x						
Li and Azarm [2002]								x		
Liang and Huang [2002]										x
Malmqvist [1994]			x							
Messac et al. [2002]							x			
Nayak et al. [2002]				x						
Nelson et al. [2001]							x			
Ong et al. [2006]	x									
Rai and Allada [2003]							x			
Simpson et al. [2001]						x				
Ulrich and Seering [1989]		x								

5.2.2.6 Agent-based method

Recently, agent-based methods have received more attentions in configuration design. Campbell *et al.* [1999] proposed an A-Design approach to explore the design space in search of the Pareto-optimal solutions. Rai and Allada [2003] developed a module-based product family design system that combined multi-agent systems, function architecturing and multi-objective optimization. Liang and Huang [2002] investigated the role of intelligent agents in satisfying customer requirements using a collaboration information system. Darr *et al.* [1994] developed the Automated Configuration-Design Service (ACDS) system to meet the requirement of concurrent engineering. ACDS is a collection of loosely coupled, autonomous agents that organize synchronous communication among themselves based on high-level specifications that a designer provides for a desired design.

The elegance of agent-based approaches is that multiple tasks can be assorted to different agents. The agents can fulfill the specified tasks using appropriate tools or algorithms, such as SA and GA. For example, the A-Design framework is a combination of GA, SA, and Tabu search. Agent-based design offers a promising solution to rapid and concurrent product configuration design [Shen *et al.*, 2001].

5.3 Multi-objective Struggle Genetic Algorithm Design Synthesis

5.3.1 *Problem formulation*

For the design synthesis problem, the design requirements are defined according to the KCs to address the customers' needs. The physical components in the component catalog are identified, as has been discussed in Chapters 3 and 4. Next, the configurations of a product or a product family are generated and evaluated with respect to multiple product performance criteria. Through these processes, product 'optimality' can be addressed as the optimal trade-offs between these multiple performance criteria.

Based on previous discussions (Section 4.5), a component catalog is denoted as $m = [s_1, s_2, ..., s_m]^T$, where s_i is the ith component slot that corresponds to a function f_i, and $s_i = [m_i^1, m_i^2, ..., m_i^{n_i}]^T$, where n_i is the number of components in the component slot. Using a more general form, the components can be collectively represented as $m = [m_1, m_2, ..., m_n]^T$, where n is the total number of physical components. Thus, the components $m_1, m_2, ..., m_n$ are considered as the DPs. Due to the constraints imposed by the compatibility among components, an arbitrary configuration may not be feasible. An optimal solution must also be feasible.

The desired performance can be specified according to the KCs, which are denoted as $k = [k_1, k_2, ..., k_p]^T$. The design specifications with respect to KCs are categorized into two types, namely the design objectives and design constraints. In a typical product design problem, these design objectives are usually non-commensurable, such as minimizing cost, minimizing manufacturing time, etc. The design constraints refer to a number of restrictions imposed on the design artifact, such as the product dimension, material, power consumption, etc. For each possible product configuration, the performance with respect to the KCs can be computed based on the properties of the components and the relationships defined in the product platform.

Let $T(m) = [t_1(m), t_2(m), ..., t_r(m)]$, where $t_1(m), t_2(m), ..., t_r(m)$ are the r objective functions. The values of $T(m)$ constitute an attribute space. Let $G(m) = [g_1(m), g_2(m), ..., g_l(m)]$, where $g_1(m), g_2(m), ..., g_l(m)$ are the l constraint functions. Both $T(m)$ and $G(m)$ can be computed from the attributes of the component m, according to the relations defined in the product platform. Assuming that the design requirements for a particular design problem are initiated as design constraints $g^0{}_1, g^0{}_2, ..., g^0{}_l$, it is required that the suitable m be set, such that the corresponding $G(m)$ is computed to fulfill the design constraints. Thus, the problem can be formulated as a MOOP (Figure 5.4).

The design problem is reduced to the selection and combination of the physical components from the existing products to minimize the design objectives, subject to the constraints. In a fully developed design database, sufficient number of physical components can be retrieved as the building elements. Thus, there will be many possible combinations. Moreover, for a typical engineering design problem, little is known about the shape and modality of the attribute space *a priori*. For example, the objective functions can be linear or nonlinear; the attribute space can be convex or non-convex, discrete or continuous. This is especially true considering a generic method applicable to the design of various types of products. A robust and efficient search and optimization method is needed to find the optimal solutions. In this book, the MOSGA method is adopted to solve this problem.

Min. $T(m) = [t_1(m), t_2(m), ..., t_r(m)]$

s.t. $m \in S$ and $[g_1(m), g_2(m), ..., g_l(m)] \le [g^0{}_1, g^0{}_2, ..., g^0{}_l]$

where:

$m = [m_1, m_2, ..., m_n]^T$ is the vector containing the physical components.

The inequality in the constraint equation applies component-wise, i.e. $g_i(m) \le g^0_i$, *for* $j = 1, 2, ..., l$.

Figure 5.4 Problem formulation of design synthesis and evaluation

5.3.2 *The MOSGA algorithm*

The MOSGA algorithm for the design synthesis problem is discussed in this section and illustrated in Figure 5.5.

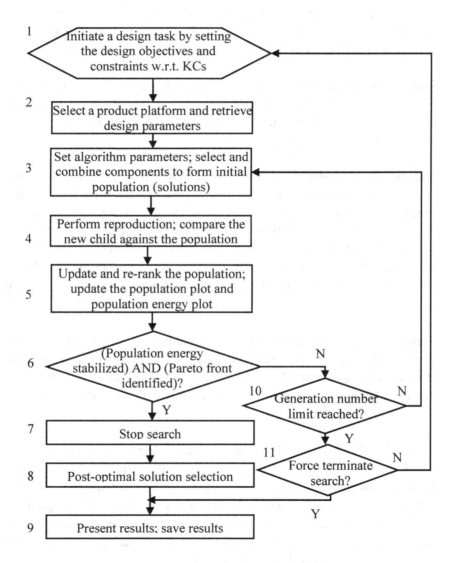

Figure 5.5 Flowchart of MOSGA for the design synthesis problem

Step 1 A design task is initiated by selecting a set of KCs and setting them as design objectives or constraints. For various types of products, the KCs are different. Hence, the design objectives and constraints are different.

Step 2 A product platform is chosen and the physical components are retrieved from the component catalog as the DPs. Retrieval of components is subject to the design constraints, which is discussed in Section 5.3.3.4.

Step 3 The parameters of the algorithm are set up. In particular, the population size and the number of generations are the most important control parameters. The initial population is generated according to the population size.

Step 4 Two individuals are randomly selected from the population, and crossover and mutation are performed to generate the offsprings. In this process, the stochastic universal selection, the single-point crossover, and the bit-wise mutation, which are common GA operators, are used to generate the offsprings. Next, the two objective functions of each child are calculated. A child is compared with every individual in the population with respect to the objective functions, through which the rank of the child is obtained.

Step 5 The algorithm searches for an individual that is most similar to this child and replaces it with the child if the child has a lower rank, or if the child dominates over it. The rank of the population is updated if the child has been inserted. The population energy and the population plot are updated accordingly.

Step 6 The Pareto-optimal frontier is identified in this step. It should be noted that the purpose of this method is to find adequate design variants, instead of identifying each and every global optimum. Therefore, once the design space has been explored sufficiently such that the population converges to a specific curve/surface, presumably the Pareto-front, the search is stopped. The mechanism to ensure a sufficient exploration of the design space is discussed in Section 5.3.3.3.

Steps 7-11 In the subsequent steps, the algorithm controls the search by checking whether the number of generations reaches a pre-defined limit, or whether the user force stops the search. The Pareto-optimal set, if it is obtained, undergoes the post-optimal selection process to arrive at the final candidate solutions. Design synthesis results are saved into the database if desired.

5.3.3 *Implementation of MOSGA in product configuration design*

5.3.3.1 The structure of a chromosome

In this book, a chromosome, *i.e.*, a solution, is a combination of the available physical components. The length of a chromosome is the total number of candidate components, *i.e.*, all the components that have been retrieved for the design synthesis. Each physical component is represented as one-bit on the chromosome string and this one-bit indicates whether that component has or has not been selected to form a solution. If a component has been selected to form a solution, the bit has a value of '1'; otherwise it has a value of '0'. This results in a string of bits indicating the selection of the physical components. An additional constraint is that only one physical component can be selected from within any one component slot. For example, each column, under the title 'Solutions' in Table 5.2 denotes one chromosome.

5.3.3.2 Measurement of similarity between individuals

In Step 4 of the MOSGA algorithm, a measurement of the similarity between the individual chromosomes is required. The similarity between two individuals is measured using the combined distances in the attribute space ($T \in R^2$) and the parameter space ($S \in R^n$). The distances are calculated as follows.

Two individuals, *a* and *b*, are denoted as:

$$\boldsymbol{m}_a = \left[m_{a1}, m_{a2}, ..., m_{an} \right]^T,$$

$$\boldsymbol{m}_b = [m_{b1}, m_{b2}, ..., m_{bn}]^T,$$

where $m_{ai} = \begin{cases} 1, & \text{if } m_i \text{ is selected to form solution } a \\ 0, & \text{otherwise} \end{cases}$.

A similar definition is adopted for m_{bi}.

The distance between a and b in the parameter space is computed as:

$$D_{IST}^S = 1 - \frac{\|\boldsymbol{m}_a, \boldsymbol{m}_b\|}{m} \tag{5.8}$$

where m is the number of component slot.

Table 5.2 Chromosome structure in design synthesis

Slot	Component	Solutions								
		1	2	3	4	5	6	7	...	N
s_1	m_1^1	0	1	0	0	0	0	0		1
	m_1^2	1	0	0	0	0	0	0		0
	m_1^3	0	0	0	1	0	0	1		0
	m_1^4	0	0	1	0	0	1	0		0
s_2	m_2^1	0	0	0	0	0	0	1		1
	m_2^2	0	0	0	0	1	0	0		0
	m_2^3	1	0	0	1	0	0	0		0
s_3	m_3^1	1	0	0	0	0	1	0		0
	m_3^2	0	0	1	0	1	0	0		0
	m_3^3	0	1	0	1	0	0	0		0
	m_3^4	0	0	0	0	0	0	0		1
⋮										
s_m	$m_m^{n_m}$	0	1	0	0	0	0	1		1

$\|m_a, m_b\|$ is the total number of components that satisfies $m_{ai} = m_{bi} = 1$, $(i = 1,...,n)$. Since only one conceptual component can be selected from within any one component slot for each member product, $0 \le \|m_a, m_b\| \le m$. It follows that $0 \le D_{IST}^S \le 1$, where '0' indicates that two individuals use the same set of components and '1' indicates that they are completely different.

The distance between *a* and *b* in the attribute space is computed as:

$$D_{IST}^T = \frac{\sqrt{\sum_{i=1}^{r}(T_i(a) - T_i(b))^2}}{[\![T]\!]} \qquad (5.9)$$

The numerator is the Euclidean distance of two individuals in the attribute space, and $[\![T]\!]$ is the diameter of the attribute space. Since the shape of the attribute space is not known *a priori*, the exact value of $[\![T]\!]$ is not available. Therefore, it is approximated as the maximum distance between two individuals in the current population, *i.e.*, $[\![T]\!] = \max\left(\sqrt{\sum_{i=1}^{r}(T_i(m_u) - T_i(m_v))^2}\right)$, where m_u and m_v are individuals in the current population. Based on this formulation, $0 \le D_{IST}^T \le 1$.

Finally, the combined distance is computed as

$$D_{IST} = \frac{1}{2}\left(D_{IST}^S + D_{IST}^T\right) \qquad (5.10)$$

$D_{ist} \in [0,1]$, and a value close to '0' indicates a higher similarity between two individuals, and a value close to '1' indicates lower similarity.

5.3.3.3 Ensuring sufficient exploration of the design space

Step 6 requires that the design space be explored sufficiently to obtain the optimal or near optimal solution. Two simple criteria are presented next. A sufficient exploration of the design space is confirmed when both criteria are satisfied.

(1) The average population energy is stabilized at a reasonably low level. The average population energy refers to the average values of the objective functions according to the current population. The energy is calculated for each generation. It will decrease gradually, and converge to a relatively stable state after a certain number of generations. It is observed that few new Pareto optima are discovered after the search reaches the stable state, and hence this is an indication that the design space has been effectively explored. A typical population energy curve is shown in Figure 5.6.

(2) The Pareto-optimal front can be identified in the population plot. A population plot illustrates the individuals as points in the attribute space, with their objective functions as the coordinates. During the GA-based search, new individuals will be added into the plot. It is observed that at the early stage of the search, a number of new individuals are produced. As the search proceeds, fewer new individuals are produced, and the non-dominant optimal solutions gradually form a specific curve/surface. When such a curve/surface becomes apparent, the search is stopped manually by the designer. A typical population plot is shown in Figure 5.7, where the diamonds represent the initial populations and the crosses represent the individuals generated during the search. The search converges to a Pareto-optimal set, indicated by the triangles.

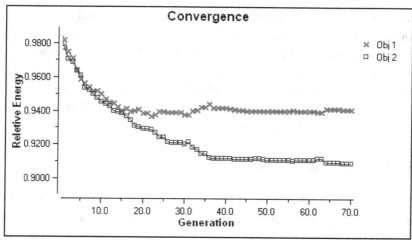

Figure 5.6 Visualization of population energy convergence

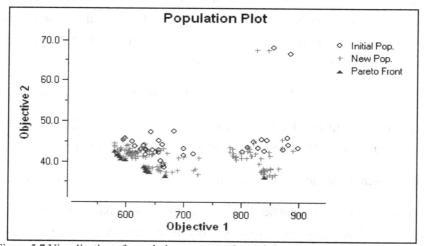

Figure 5.7 Visualization of population energy and population plot

A design synthesis system has been developed, where the population energy and population plot can be visualized. The user can observe the search process, stop the search, or fine-tune the parameters with a new iteration. For parameters fine-tuning, two parameters are essential, namely the population size and the number of generations. A user can carry out a few tentative design synthesis sessions to determine these parameters.

Generally, the design space can be sufficiently explored with a large population size and a large number of generations. However, this will significantly increase the computational load. A practical method is to start from a relatively small population size and a small number of generations, and increase them gradually. The parameters are finalized when the increase of these two parameters does not result in an obvious improvement of the Pareto-front.

5.3.3.4 Applying design constraints

The design constraints are used to filter the physical components so that only those components that can satisfy the constraints are used in the actual design synthesis. The product case from which a component is extracted from is called its host product. A filtering process is carried out through a comparison of the KCs of the host products with the specifications of a new design task. It is assumed that the design specifications of a new design task constitute a design range vector $R_d = \left[d_1, d_2, ..., d_q\right]^T$. R_d is defined with respect to a set of KCs (k_i, $i=1,...,q$). The subscript here is q instead of p as not all the KCs are used as the design constraints. A host product is characterized by the same set of KCs, which determines a system range vector $R_s = \left[s_1, s_2, ..., s_q\right]^T$. Thus, a pair-wise comparison can be made between R_s and R_d, based on the type of KCs. This leads to a set of results showing whether a system range can satisfy a design range.

At the same time, each physical component m_j, ($j=1,...,n$) is related to a few KCs through the transformation matrices TR_2 and TR_3, i.e., there exists a subset of $\{k_i \mid i=1,...,q\}$, which elements are the KCs related to m_j ($j=1,...,n$). If, for every element in this subset, the corresponding s_i satisfies d_i, the candidate component c_j will be selected for the design synthesis. Otherwise, the component is discarded. This process is shown in Figure 5.8.

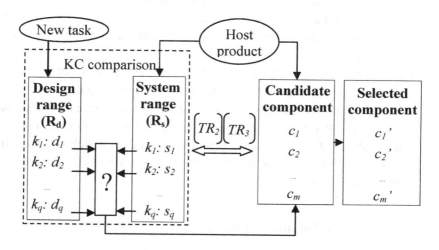

Figure 5.8 Retrieving components through KCs comparison

5.3.4 Precautions and limitations

The MOSGA method is a generic method for design synthesis based on a pre-defined product platform and component catalog. The following observations can be made concerning the application of this method.

(1) The method is effective when the product has a modular structure. This requirement arises from the need to formulate the design synthesis as a configuration design problem. For products that have an integral architecture, many product functions are coupled, and hence, it may be difficult to identify reusable components and deal with the incompatibilities among the reused components. Therefore, the method is not applicable to products with an integral architecture. Fortunately, a modular structure has been adopted in many products, such as home electronics, power tools, automobiles, etc. Thus, the design reuse method is not restricted to specific products.

(2) For convenience of narration, some of the discussions are presented in the context of designing a single product, such as the structure of a chromosome and the measurement of the similarity between the individuals. However, the MOSGA method is applicable to product family design as well. To do so, the structure of chromosome and the similarity measurement of the individuals need to be modified slightly.

(3) Another limitation is that the MOSGA algorithm does not guarantee that the best solution is found. The method is robust in the sense that it is able to explore the design space effectively and arrive at near global optimum. However, being a heuristic algorithm, the method does not guarantee that all the global optimums are found, which is especially true for a product design problem that involves significant complexity and nonlinearity.

5.4 Post-optimal Solution Selection

The multi-objective optimization methods will result in a Pareto-optimal set. Usually, a designer needs to select one or a few candidate solutions to be further developed in the subsequent design processes. A few classical post-optimal solution selection methods have been reported, such as the compromise programming method, pseudo-weight vector method, and marginal rate of substitute method [Deb, 2001]. These methods have their merits in various aspects. The bottom-line is some meta knowledge is needed to select an optimal solution from the Pareto set. A designer needs to indicate his/her preference of the design objectives through assigning different weights to them, thus leading to configurations with different priorities with respect to different objectives.

In this book, a relative weight is assigned to each objective and a weighted fitness $T_w(\boldsymbol{m})$ is obtained for every Pareto-optimal solution.

$$T_w(\boldsymbol{m}) = \sum_{i=1}^{r} \frac{w_i \left(t_i(\boldsymbol{m}) - t_i^{\min}\right)}{t_i^{\max} - t_i^{\min}} \quad (5.11)$$

where r is the number of objectives; w_i is the weight assigned to the ith objective and $\sum_{i=1}^{r} w_i = 1$; and t_i^{max} and t_i^{min} are the maximum and minimum objective functions for the ith objective, respectively.

Thus, $(t_1^{min}, t_2^{min}, \ldots, t_r^{min})$ constitutes the coordinates of a utopian ideal solution. Hence, $T_w(m)$ is a normalized, weighted distance that a solution is to the ideal solution. The final solution is chosen as the one(s) with the smallest weighted fitness. If the weighted fitness values of several solutions are very close, additional meta-knowledge is required to assess the merits of the candidate solutions. Figure 5.9 illustrates a Pareto-front based on two objectives, where each black square denotes a solution of a product.

A further remark can be made concerning the post-optimal solution selection process and the traditional methods, such as the weighted sum method which convert a MOOP into a SOOP. Since both methods involve user-defined weights to search for the optimal solutions, it is important to note the difference between them, and the merits of multi-objective optimization when it is much more complicated to be solved.

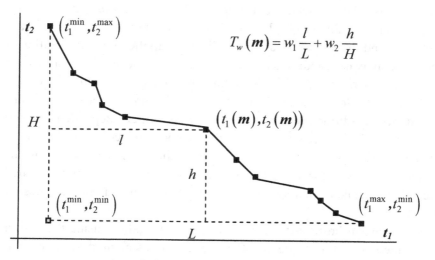

Figure 5.9 Pareto-front and post-optimal solution selection

The multi-objective optimization actually involve two steps, namely (1) searching for the Pareto-front that reveals multiple trade-offs between the objectives, and (2) choosing one optimal solution within this Pareto-front using higher-level preference information. Thus, it is dealing with two search spaces, namely the parameter space and the attribute space. In contrast, the single-objective optimization combines these two steps into a single step, and it deals with only one space, namely the parameter space. Consequently, the Pareto-optimality information may not be revealed in this one-step process. Such information may be helpful for a designer to make decisions when choosing the solutions from the Pareto-front. Thus, the MOOP is advantageous to SOOP in providing an insight to the optimization problem and the search spaces.

5.5 A Case Study

This section presents a case study to design a fan filter unit (FFU) using the design synthesis approach presented in this chapter. The FFU is the key device in clean room products. It draws in air from the outdoor space, removes the unwanted particles, and supplies clean and laminar airflow continuously into a clean room. Figure 5.10 illustrates the structure of a horizontally mounted FFU. The performance requirements of the FFU are rigorous and diverse with respect to different applications. In addition, a customer may have individual demands, such as a special size or a specific material. Therefore, the FFU manufacturers have to be able to provide customizable, low-cost products, while conforming to various industry standards. Traditionally, the FFU design is largely dependent on the expertise of the designers, with the support of computer-based or paper-based design catalog. In comparison, the method presented in this book is based on a formalized design reuse framework, with an automated development and evaluation of the solution alternatives at the early design stage.

The aim of the case study is to design FFU products using the design reuse method and illustrate the effectiveness of this method by comparing it with the traditional experience-based method. Parallel design sessions were carried out by two groups of designers, with respect to the same set of design requirements shown in Table 5.3. Group A carried out the design based on experience. Group B performed the task using the design reuse system. Both groups consist of two designers, namely a product planner who proposes the configuration of the product, and a technical engineer who is responsible for estimating the technical feasibilities. The designers in these two groups have the same level of experience, *i.e.*, the average number of years in the profession is equivalent (3.5 years). The product planner in Group B was given a short training course on the use of the design reuse system. In addition, the subjective weights assigned to the design objectives would influence the results. Group B adopted quantitative weights, namely $w_1 = 0.5$ for cost, $w_2 = 0.4$ for manufacturing time, and $w_3 = 0.1$ for the product weight. A similar strategy was adopted by Group A. However, estimated values of the objective functions were used where the exact values were not available. The procedures using different approaches are presented next, and the outcome is compared with respect to solution variety and solution superiority.

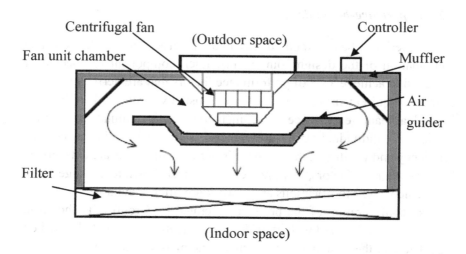

Figure 5.10 FFU structure and major components

Table 5.3 Design objectives and design constraints

Objectives	
1) Minimize cost 2) Minimize manufacturing time 3) Minimize weight	
Constraints	
motor type	AC-1PH
air quantity	$1375 m^3/h \pm 10\%$
air velocity	$0.42 m/s \pm 10\%$
air uniformity	$\leq 15\%$
service cleanliness	Class1000
noise level	$\leq 50 dBA$
casing material	Aluminium
casing size	3×4
Vibration	$<0.025G$
mounting type	Horizontal

5.5.1 Experience-based design

The experience-based method involves a procedural selection and combination of the design components. Each component is selected or designed to achieve a set of performance factors that are derived from the design requirements. The designer usually relies heavily on individual/team experience, documentation and simulation results. Therefore, individual/team preferences significantly influence the efficiency and validity of the design. Since the components are determined by a number of factors, it is not easy for a designer to manage them simultaneously. Modifications and rework are often inevitable.

The configuration of the product generated by Group A is shown in Table 5.4. The cost and weight of the components are estimated based on past data and the designers' experience. The manufacturing time was not estimated according to current design practices. It is noted that the division of the design components is different for the experience-based method and the design reuse method. For the convenience of comparison, this book has purposely rearranged the components into an identical format.

Table 5.4 Configurations of the candidate solutions based on two methods

Component	Item	Solution by Group A	Solution by Group B
Blower	Type	R4E 310-AP20-01	R4E 310-AF12-05
	Cost (S$)	130	110
	Weight (kg)	4	3.8
	Manufacturing time(min)	–	0
Control	Type	LSC-S1	Motor Integrated
	Cost (S$)	10	0
	Weight (kg)	0	0
	Manufacturing time(min)	–	0
Filter	Type	HEPA filter	(2*4)HEPA T70
	Cost (S$)	200	180
	Weight (kg)	10	10
	Manufacturing time(min)	–	0
Casing	Material	Aluminium	Aluminium
	Size (H*W*L) (mm)	330*870*1170	325*870*1170
	Cost (S$)	220	184
	Weight (kg)	25	23
	Manufacturing time(min)	–	18.4
Air guilder	Type	set-piece	Set air guiders
	Cost (S$)	15	18
	Weight (kg)	2	3.5
	Manufacturing time(min)	–	2
Inlet cone	Shape	Square	Circular
	Type	punched hole	Aluminium coil set
	Cost (S$)	10	24
	Weight (kg)	2	0.2
	Manufacturing time(min)	–	1.4
Insulation	Type	corner blocks	Corner block
	Cost (S$)	5	8
	Weight (kg)	2	3.8
	Manufacturing time(min)	–	3
Summary	Cost (S$)	590	524
	Weight (kg)	45	40.9
	Manufacturing time (min)	approx. 30	28.2

5.5.2 Product design using the design reuse approach

In this case study, 22 FFU product cases have been collected, modeled and stored in the database. Product analysis has been carried out, where the SOM method was used to generate the feature map (Figure 5.11(a)). This feature map further facilitates the establishment of the FPA, which is composed of seven common functions (Figure 5.11(b)). A set of 18 KCs is formalized and seven common functions which are the correlation of KCs are established as TR_2 (refer to Table 5.5). A component catalog was built to include all the unique conceptual components according to product function (Table 5.6).

Following the procedures presented in Section 5.3, design synthesis is carried out using the design reuse prototype system. The major steps are:
Step 1 Input the design objectives and design constraints as shown in Table 5.3.
Step 2 Select a product platform and retrieve the physical components (Figure 5.12). In this step, only the components that satisfy the design constraints are selected.
Step 3 Set the parameters of the algorithm, namely, the population size and the number of generations. To determine the suitable algorithm parameters, a few tentative synthesis sessions are performed. A population size of 60 and number of generations of 100 are adopted for this case study.
Step 4 Identify the Pareto-front by observing the population plot and population energy.
Step 5 Set a relative weight to each objective. Calculate the weighted fitness and present the solution candidates (Figure 5.13).

Based on the above procedures, the design synthesis has arrived at a Pareto-optimal set of 33 solution variants. Each solution presents a possible configuration of the product. An excerpt of the solution configuration is shown in Figure 5.13. In this figure, each cell shows the name of the component. The numbers in the brackets are the indices of the product and the component. A list of the objective functions of the solutions is shown in Figure 5.14.

Optimization in Product Design 147

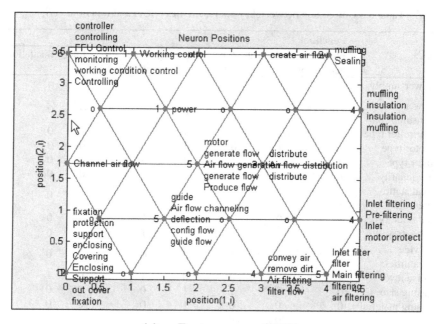

(a) Feature map of FFU

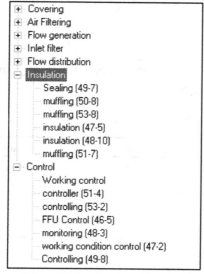

(b) FPA of FFU products

Figure 5.11 Establishment of function-based product architecture using SOM

148　Design Reuse in Product Development Modeling, Analysis and Optimization

Table 5.5 Correlation between key characteristics (KCs) and functions (TR_2)

KCs	Functions						
	Flow generation	Control	Air filtering	Casing	Flow distribution	Inlet	Insulation
power supply	1*	1	0	0	0	0	0
power consumption	1	1	1	1	0	0	0
controller type	0	1	0	0	0	1	0
motor type	1	1	0	0	1	0	0
filter type	0	0	1	0	0	0	0
air quantity	1	0	1	1	1	0	0
total static pressure	1	0	1	1	1	0	0
air velocity	1	0	1	1	1	0	0
air uniformity	0	0	1	1	1	0	0
service cleanliness	0	0	1	0	0	0	0
noise level	1	0	0	0	1	0	1
casing size	0	0	1	1	0	0	0
casing material	0	0	0	1	0	0	0
vibration	1	0	0	1	0	0	1
mounting type	1	0	0	1	1	1	0
cost	1	1	1	1	1	1	1
weight	1	1	1	1	1	1	1
manufacturing time	1	1	1	1	1	1	1

*1 indicates that a KC is dependent on a function; 0 indicates that a KC is not dependent on a function.

Table 5.6 Function and component slot

Function		Component slot		Number of components
f_1	Flow generation	s_1	Motor/Blower	11
f_2	Control	s_2	Controller	3
f_3	Air filtering	s_3	Filter	9
f_4	Casing	s_4	Casing	11
f_5	Inlet cone	s_5	Motor support/Inlet cone	6
f_6	Flow distribution	s_6	Air guiders	3
f_7	Insulation	s_7	Muffler	3

Optimization in Product Design   149

Index	Available Platforms			Create time			
	14 Fan Filter Unit			2/16/2005			

ID	Component Name	Cost	Weight	Production Time	Pro...	Product Name	Functio...	Detail
	6 SUS Casing	284	18.3999996	25		46 FFU-0501-SG		3 standard SUS Casing
	7 N.A.	0	0	0		46 FFU-0501-SG		5 3PH power supply. No speed c
	5 Punch square array	18	3.5	2		46 FFU-0501-SG		8 Inlet cone
	4 ULPA	240	10	0		46 FFU-0501-SG		9 Ultra Low penetration air filter
	3 R4D-310-APA0-09	130	3.79999995	0		46 FFU-0501-SG		10 OEM German ebm papst
	1 Set air guiders	18	3.5	2		46 FFU-0501-SG		11 standard 3 planer air guider
	2 slanted air guider	18	3.5	2		46 FFU-0501-SG		12
	7 Integrated controller	10	0.5	0		47 FFU-0502-SG		2
	4 Rockwool Insulation	10.5	4.80000019	3.5		47 FFU-0502-SG		5
	1 4-4HEPA	320	20	0		47 FFU-0502-SG		7 Thickness 70mm HEPA filter
	6 Motor guard	27	4	2.5		47 FFU-0502-SG		8
	5 4-4 Galvalume casing	424	40.4000015	41.5		47 FFU-0502-SG		9
	3 R3G 400-AD36-61	390	5	0		47 FFU-0502-SG		10 OEM size 400 motor by ebm pa
	2 Deflectors	21	6.69999980	9		47 FFU-0502-SG		11 standard 3 set air deflector
	7 R4D 310-A026-09	110	3.79999995	0		48 FFU-0401-TW		4 OEM 310 AC type motor, ebm
	6 2-4 HEPA	180	10	0		48 FFU-0401-TW		5 High efficiency particulate air fil
	4 Galvalume casing	300	21.2000007	28.5		48 FFU-0401-TW		6
	1 standard Inlet cone	28	3.5	1.79999995231		48 FFU-0401-TW		8
	2 Air flow guider	10	4.40000009	8		48 FFU-0401-TW		9
	5 Rockwool insulation	8	3.79999995	4.19999980926		48 FFU-0401-TW		10
	3 AL shaft fixation	300	21.2000007	28.5		48 FFU-0401-TW		11

Figure 5.12 Retrieve product platform and conceptual components

	Candidate Solution A	Candidate Solution B	Candidate Solution C
Component 1	Al shaft (51-4)	Al shaft (51-4)	Al shaft (51-4)
Component 2	2-4 HEPA T70(71-4)	hepa t75(77-7)	hepa t75(77-7)
Component 3	R4E 310-AF-12-05(53-3)	R4E 310-AF-12-05(53-3)	R4E 310-AF-12-05(53-3)
Component 4	Set air guiders(46-1)	slanted air guider(46-2)	Set air guiders(46-1)
Component 5	RockWool(67-3)	RW STD Muffler(66-3)	RW STD Muffler(66-3)
Component 6	Coil Al(74-2)	STD-310(75-6)	STD-310(75-6)
Component 7	N.A.(46-7)	N.A.(46-7)	N.A.(46-7)
Objectives	Cost; Weight; Production time;	Cost; Weight; Production time;	Cost; Weight; Production time;
Obj. Values	524.00 40.90 28.20	526.00 40.80 28.40	526.00 40.80 28.40
Weighted fitness	0.08	0.09	0.09

Figure 5.13 Configurations of the Pareto-optimal solutions and evaluation results

150 *Design Reuse in Product Development Modeling, Analysis and Optimization*

Number of Pareto optima: 33								
	Cost	Weight	Production	Weighted	* Solution 16 *522.00	40.70	33.40	0.36
	(0.5)	(0.1)	time(0.4)	fitness	* Solution 17 *727.00	40.20	29.70	0.46
					* Solution 18 *726.00	40.40	28.40	0.39
* Solution 1 *520.00		40.80	33.20	0.35	* Solution 19 *527.00	40.60	29.70	0.16
* Solution 2 *724.00		40.50	28.20	0.37	* Solution 20 *644.00	36.90	30.20	0.32
(Solution 3 *526.00)		40.80	28.40	0.09	* Solution 21 *632.00	40.10	35.00	0.61
* Solution 4 *644.00		36.90	30.20	0.32	* Solution 22 *637.10	37.70	34.20	0.54
* Solution 5 *646.00		36.80	30.40	0.33	* Solution 23 *517.10	41.70	32.20	0.30
* Solution 6 *522.00		40.70	33.40	0.36	* Solution 24 *726.00	40.40	28.40	0.39
* Solution 7 *640.00		39.20	30.50	0.36	* Solution 25 *517.10	41.70	32.20	0.30
* Solution 8 *635.10		37.80	34.00	0.53	* Solution 26 *646.00	36.80	30.40	0.33
* Solution 9 *840.00		36.40	35.20	0.88	* Solution 27 *511.10	44.10	32.30	0.33
* Solution 10 *642.00		36.70	35.40	0.60	* Solution 28 *637.10	37.70	34.20	0.54
* Solution 11 *844.00		36.50	30.20	0.61	* Solution 29 *646.00	36.80	30.40	0.33
* Solution 12 *646.00		36.80	30.40	0.33	* Solution 30 *515.10	41.80	32.00	0.29
(Solution 13 *526.00)		40.80	28.40	0.09	* Solution 31 *647.00	36.60	31.70	0.40
(Solution 14 *524.00)		40.90	28.20	0.08	* Solution 32 *844.00	36.50	30.20	0.61
* Solution 15 *644.00		36.90	30.20	0.32	* Solution 33 *640.00	36.80	35.20	0.59

Figure 5.14 The objective functions of the Pareto-optimal set

Finally, to choose the candidate solutions from the Pareto-optimal set, weights are assigned to the objectives as cost ($w_1 = 0.5$), manufacturing time ($w_2 = 0.4$) and weight ($w_3 = 0.1$). These weights are assigned according to the designer's knowledge of the importance of these objectives. It can be observed that the weighted fitness of the three solutions are very close, namely Solution 3- *0.09*, Solution 13- *0.09*, and Solution 14- *0.08*. A closer study of the components in the three solutions is required. It is observed that the *Al-shaft (51-4)* has been shared by the three solutions corresponding to component 1 (casing) (Figure 5.13). Similarly, identical components have been used for components 3 (blower) and 7 (controller), respectively. For the other components, namely 2, 4, 5, and 6, there are only minor differences among the attributes of the actual components. For example, Solution 14 has used *2-4 HEPA T70 (71-4)* as the filter, while solutions 3 and 13 have used *HEPA T75 (77-7)*. Differences exist concerning the filter thickness, cost and weight. However, these differences are insignificant. Therefore, the three solutions are almost identical to each other. The designer can choose any one without making much difference in the final results. The product configuration generated by Group B is shown in Table 5.4.

5.5.3 *Comparison of the two methods*

The effectiveness of the experience-based method and the design reuse method are compared with respect to solution variety and solution superiority.

The design reuse method has generated more design variants. The design synthesis based on multi-objective optimization has generated 33 Pareto-optimal solutions. Post-optimal selection leads to three solutions from the Pareto set. Although only one solution is chosen for further development, it is advantageous to have a few backups, in case demerits are found in the chosen solution in the subsequent design process.

The design reuse method can generate solutions that are superior in the design objectives, namely cost, weight and manufacturing time. The final adopted solution has the objective functions of cost (S$) - *524*, weight (kg) - *40.9* and manufacturing time (min) - *28.2*. In comparison, the solution generated by Group A is estimated as cost - *590*, weight - *45* and manufacturing time - *30*. There is a notable improvement in the design objectives. This can be attributed to the advantage of the design reuse method, namely the GA-based search has explored the design space more effectively to reuse the best fitted components.

5.6 Summary

Optimization is considered as a major enabler of automated design synthesis. This chapter has introduced the configuration design problem and its relationship with engineering optimization. A few prevalent approaches are introduced, such as the weighted sum approach, GP, multi-level programming and evolutionary algorithms. GA is especially suitable for configuration design, and hence is discussed in greater detail.

The product configuration design problem is defined and the commonly used design synthesis approaches are discussed. A GA-based optimization algorithm, namely MOSGA, is used to support automated design synthesis and evaluation, which caters to the multiple design objectives and constraints. A comprehensive case study of the design of FFU is presented using the design reuse method, where the design reuse method is compared with the traditional experience-based design method. It is demonstrated that the optimization method can effectively explore the design space and generate the Pareto-optimal solutions. The automated design synthesis has outperformed the experience-based method with respect to the major criteria of product performance.

Chapter 6

Cost Estimation in Product Development

In design synthesis, a set of candidate solutions are generated which superiority must be evaluated according to the selected criteria. In Chapter 5, solutions have been evaluated with respect to the KCs that are set as the design objectives and constraints. In the subsequent discussions, two critical criteria are used to evaluate the candidate solutions, namely the product cost and the product performance. This chapter focuses on the evaluation of product cost. After an introduction of the importance and requirements of cost estimation (Section 6.1), a mixture of cost elements is discussed in Section 6.2 and a few cost modeling techniques are discussed in the context of product development. In Section 6.3, cost estimation in product family design is investigated according to different cost measurements. Finally in Section 6.4, an empirical cost model based on design reuse is proposed. It should be noted that this chapter is not aimed at developing a conclusive cost modeling method because the complexity and dynamic nature of the problem would make such an effort inappropriate.

6.1 Introduction

Despite the many unforeseeable and uncontrollable factors, most corporations choose to limit or promote their business goals based on thorough profit plans and budgets that manage the entire spectrum of production/service processes. While profit is the ultimate gauge of business success (although many corporations do not claim so), cost is a key factor to be considered in such an effort. Its importance can be seen from a simple observation:

$$\text{Profit} = \text{Price} - \text{Cost}$$

Thus, for any manufacturers, cost estimation and control is indispensable for successful business management.

From the perspective of product development, cost estimation at the planning and design stage is of special interest. This is because 70 – 80% of the production cost is committed at this stage, whereas the cost incurred at this stage accounts for only 7% of the production cost [Ehrlenspiel, 1985]. Figure 6.1 illustrates the cost incurred and committed in different production activities. It is apparent that wise decisions at the early design stage are more effective at reducing cost and hence can increase the chances of business success.

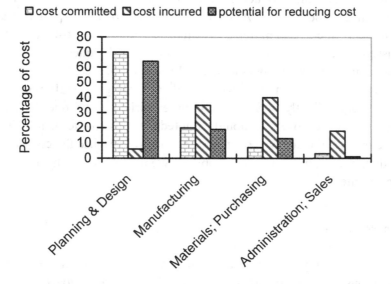

Figure 6.1 Product cost incurred and committed in different production activities [Ehrlenspiel, 1985]

However, cost estimation is not easy at the early stage of product development. First, there are many factors to be considered. For example, cost is related to different departments and manufacturing processes. Hence, cost must be properly allocated among the various production activities, such as market investigation, research and development (R&D), manufacturing, service, administration, etc. Second, information for decision support is highly uncertain at this stage. For example, the colors and materials of a product preferred by the customers cannot be predicted accurately. The change of color or material as a request from the customers may cause a series of adjustments in the design, process planning, purchase order, and inventory stages, which have considerable effects on the production cost. In addition, considering the ever-growing collaborations among different departments and corporations, various cost elements are highly interrelated and should not be considered independently. This factor has made cost estimation more complicated.

Therefore, cost estimation requires a thorough analysis of the cost factors, a proper allocation of the cost elements and systematic procedures to perform estimation. Section 6.2 presents an extensive study of the cost structures and cost modeling techniques to address these issues.

6.2 Product Development Cost

To develop a comprehensive cost model, there are two essential and interrelated factors, namely the cost structure and the cost modeling method. The cost structure refers to the constitutive elements of production cost. The cost modeling method is the scheme by which these cost elements are aggregated into a logical formulation. In this section, these two factors are discussed in the context of product development.

6.2.1 *Cost structure*

Basically, the cost elements can be considered from two perspectives [Michaels and Wood, 1989], namely the top-down perspective and the bottom-up perspective.

(1) The top-down perspective emphasizes *what the customers are buying in*, *e.g.*, speed, range, reliability, etc. Accordingly, cost estimation is based on various high-level product utilities, such as product functions and performance characteristics. Measurements of these elements include the design specifications and the customer's satisfaction. Statistical techniques are constantly used to derive the cost.
(2) The bottom-up perspective emphasizes *what the manufactures are paying for*, *e.g.*, labor, material, sub-contracts, and overhead. Cost estimation is based on the basic manufacturing elements, such as material, labor, energy consumption, etc. Measurements of these elements can be time, operation hour, volume, weight, etc. Cost is derived from the multiplication of unit price of the elements and quantity of usage of such elements.

The bottom-up perspective can facilitate more accurate estimations given that sufficient information can be attained from an analysis of the production operations. Thus, much effort has been carried out in the industry and the academia to classify the basic manufacturing elements into logical types.

Total manufacturing costs can be divided into direct costs and indirect costs (overhead) [Hundal, 1997]. Direct costs refer to the costs that can be assigned to the products or operations, such as material and direct labor. Indirect costs refer to the costs that cannot be clearly associated with a particular product or operation, and must be apportioned among all the cost units. While the allocation and estimation of direct costs is relatively easy, estimation of the overhead is complicated especially when the costs are associated with multiple departments of an organization.

Costs can also be divided into variable costs and fixed costs according to whether the cost is dependent on production volume [Shuford, 1995; Hundal, 1997]. Variable costs can be derived from a product being manufactured or the service being rendered, *e.g.*, materials incorporated into the product or labor-hours related to the operations. The production volume determines the amount of material and labor-hours needed. Fixed costs refer to the portions of costs that are invariant to changes in the production volume and that remain constant over a period of time.

Examples of fixed costs include the real estate taxes and administrative salaries. It should be noted that there is no clear-cut distinction between fixed cost and variable cost. In a dynamic manufacturing environment, variation in cost should always be a consideration, such that no cost is perfectly fixed.

Another important factor in cost estimation is the *learning effect*. Learning effect is used to describe the reduction in manufacturing time and cost when similar products are produced repeatedly. It can be modeled as a learning curve and expressed as Equation (6.1) [Delionback, 1995].

$$c = c_1 N^r \qquad (6.1)$$

where c is the time or cost per cycle or unit; c_1 is the time or cost for the first unit of the cycle; N is the number of cycles or units; r is an exponent function, which can be expressed as $r = \dfrac{\log(P/100)}{\log 2}$, where $P\%$ is the slope of the learning curve. Table 6.1 shows the correspondence between the exponent r and the slope $P\%$. The slope of the learning curve depends on many factors, such as the learning habit of individuals, the characteristics of the operations, and the improvements in production methods.

Table 6.1 Learning curve exponent r and slope $P\%$

Slope ($P\%$)	Exponent (r)
100	0
95	−0.074
90	−0.152
85	−0.234
80	−0.322
75	−0.415
70	−0.515
65	−0.621
60	−0.737
55	−0.862

The data for developing a learning curve can be collected from observations and the historical records of a given task. A learning curve can be plotted using regression analysis. When plotted on a logarithmic graph, the learning curve approximates a straight line (Figure 6.2). The learning curve can be used to forecast the production costs for a variety of tasks ranging from pure hands-on operations to pure mental work. The learning curve slope is different for different industries and for different production practices. Table 6.2 summarizes the approximate values of the learning curve slope in typical industries and production practices [Delionback, 1995].

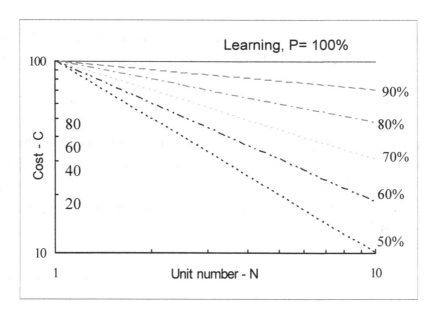

Figure 6.2 Learning curves on logarithmic coordinates

6.2.2 Cost modeling techniques

A cost modeling technique can be considered as an aggregation method to combine the elements in the cost structure. Using a mathematical format, cost can be computed as in Equation (6.2).

$$C = a_0 + F(D) \tag{6.2}$$

where C is the production cost; D is a vector of the DPs that drives cost, and a_0 is a constant. The DPs may include one or a few cost elements discussed earlier. F is the aggregation function that relates the DPs to the production cost. The aggregation function (F) may come from subjective knowledge, and/or from statistical analysis of prior production data. Several representative cost models are discussed next. These models are classified into two groups based on the cost elements and are considered from the top-down and bottom-up perspectives.

Table 6.2 Learning curve in different industries and practices

Industry or practice	Slope ($P\%$)
Aerospace	85%
Shipbuilding	80–85%
Complex machine tools for new models	75–85%
Repetitive electronics manufacturing	90–95%
Repetitive machining or punch-press operations	90–95%
Repetitive clerical operations	75–85%
Repetitive welding operations	90%
Construction operations	70–90%
Raw materials	93–96%
Purchased parts	85–88%

6.2.2.1 Expert opinion

No physical cost model is explicitly defined in this method. Cost is estimated based on experts' judgment with subjective assumptions [Michaels and Wood, 1989]. Although questionable, expert opinion is the only means to the designers when no backup or historical data is available for the design. This method is applied mostly during the product planning and conceptual design stage. Apparently, the estimation is subject to bias. Moreover, the accuracy and reliability of the estimation is degraded by the complexity of a program. A possible compensation for this weakness is to invoke opinions from multiple experts, especially those with different professional backgrounds.

6.2.2.2 Analogy

In this approach, the cost of a new project is estimated based on the costs of previous similar projects [Michaels and Wood, 1989]. The similarity of these projects can be recognized in terms of the expected outcome, available equipments and the external environment. This method is relatively simple and inexpensive. It can also provide reasonably accurate estimation results given that the similarity between projects is relevant. The limitations with this method include (1) it requires analogous products or systems, (2) the technology advancement must not be radical, and (3) it is usually applied in the same enterprise.

6.2.2.3 Function costing

In function costing, cost is attributed to the product functions. In other words, the DPs used for cost estimation are the product functions/sub-functions. This method is suitable for cost estimation at the conceptual design stage. It allows the comparison of the cost/benefit ratios of different functions. Function costing constantly resorts to a function-cost allocation table to identify the cost in correspondence with each product components (Table 6.3).

Table 6.3 Function cost allocation

Function	Part	Percent of part cost for function	Part cost ($)	Function cost $	Total
F_1	P_1	50%	5000	2500	
	P_2	20%	380	78	3418
	P_m	40%	2100	840	
F_2	P_2	30%	380	114	216
	P_k	10%	1020	102	
...					
F_n	P_2	50%	380	190	
	P_k	90%	1020	918	1318
	P_m	10%	2100	210	

A number of costing models have been developed based on the cost elements considered in the bottom-up perspective. Detailed information of the designed model, material, labor, process plan, and equipment, are necessary in these cost models.

6.2.2.4 Parametric costing

In parametric costing, cost is formulated as a function of one or more independent DPs, such as weight or the number of parts, the number of operating locations, unit power cost, etc. Parametric cost equations are also called cost estimating relationships (CERs). Usually, statistical techniques, such as regression analysis, are used to establish CERs. The model can take one of the following forms based on the application area [Michaels and Wood, 1989]. In these expressions, C denotes the cost and A is a constant. d_i is a design parameter and a_i is the coefficient or exponent that can be determined using statistical methods.

- Linear model: $C = A\sum_i a_i d_i$

- Generalized liner model: $C = A\sum_i a_i g_i(D)$, where $g_i(D)$ can be a function of various forms, such as polynomial, exponential, and sinusoidal.

- Product-exponential model: $C = A\prod_i d_i^{a_i}$

6.2.2.5 Productive hour costing

This method is useful for dealing with a large portion of the indirect costs (overhead) in production [Ostwald and McLaren, 2004]. In this method, indirect costs are first converted to equivalent hours. Next, the productive hour cost rate is determined by allocating the indirect costs to each hour of manufacturing processing. The overall indirect cost is the multiplication of these items.

162 *Design Reuse in Product Development Modeling, Analysis and Optimization*

The advantage of this method is that it has good accuracy because it considers the differences of the machine rates in different operations. Thus it combines the machine hour rate with the gross hourly rate. In addition, this method is applicable to many production systems, such as manufacturing cell, programmable automation, and machining centers.

6.2.2.6 Magnitude-based costing

Magnitude-based costing (MBC) emphasizes the degree of influence that the design components may exert on the cost based on their categories [Hundal, 1997]. First, costs are assigned to product components. Next, these components are divided into three grades according to specific properties, such as weight and cost. A higher grade has a larger impact on the overall cost, *e.g.*, A–highest, B–middle and C–Lowest. Thus, the importance of the components in the cost estimation is determined. The composition of the cost can be analyzed based on the components that can influence the cost most. Thus, the designer can focus on the components that have the most potential to reduce cost.

6.2.2.7 Activity-based costing

Activity-based costing (ABC) has been advocated by many researchers as a comprehensive method to deal with the increasing overhead expenses in a dynamic manufacturing environment [Cooper and Kaplan, 1991; Shuford, 1995; Hundal, 1997; Ostwald and McLaren, 2004]. ABC differs from the traditional cost estimation systems in the way the indirect costs are dealt with. In traditional cost systems, indirect costs are allocated to individual products, and production cost is estimated based on the production volume. In ABC, indirect costs are allocated to activities, and the production cost is computed from the aggregation of these activities. Traditional cost systems is suitable for mass production where resources (*e.g.*, direct labor, materials, processing time, space, energy, etc.) are consumed in proportion to the number of units produced. In contrast, the new manufacturing paradigm, namely mass customization, favors low volume, and a high diversity of products. In mass customization, indirect costs become a dominant part of the total production cost, and a volume-based estimation causes significant distortions. ABC can alleviate this problem by using activities, instead of production volume, as the cost drivers.

The structure and working principle of an ABC system is shown in Figure 6.3. ABC involves two basic stages, namely allocation and estimation.

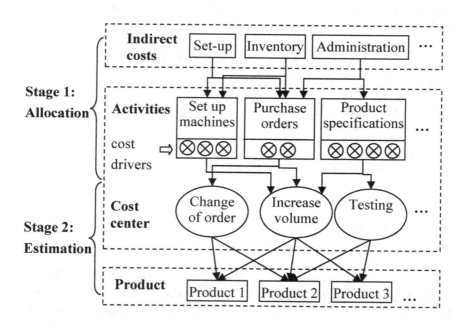

Figure 6.3 Structure of an ABC system

Stage 1 – Allocation. Indirect costs are allocated to a number of activities, which are further grouped as a set of cost centers. The indirect costs include all the resources that are not directly related to the production volume, but are necessary to support the functioning of the production, *e.g.*, the cost or time required for set-up, supporting direct labor, inventory maintenance, and administration. The resources are allocated to activities which magnitudes are determined by a set of cost drivers. A cost driver can be represented as the occurrence of a cost related to an activity. Typical cost drivers under ABC systems are listed in Table

6.4. Next, the costs are assigned to different cost centers (also called cost objects), which are invoked as a request of customers, vendors and the manufacturers.

Stage 2 – Estimation. Costs are attributed to products based on the usage of the cost centers. Each product is related to a set of cost centers. Hence, the cost of a particular product can be derived from the summation of the cost centers.

Table 6.4 Cost drivers under ABC system

Number of machine hours
Number of tests
Number of orders received
Number of different materials used
Number of material moves
Labor hours
Number of units produced
Number of setups
Number of new components
Number of vendors
Number of scheduled batches
Space occupied

Activities are the core of the ABC systems. Based on their roles in a production system, activities can be categorized into four levels in a hierarchical order, namely, the unit-level, batch-level, product-sustaining level and facility-sustaining level [Cooper and Kaplan, 1991]. Table 6.5 shows the definition and content of the activities in different levels.

The hierarchical structure of the cost activity analysis shows that many expenses committed to the products are not necessarily proportional to the number of units of the products being produced. This implies that, within a specific production session, these expenses will not increase/decrease when the production volume grows/drops. In traditional cost systems, these expenses are considered as fixed costs. However, such a practice usually leads to a distorted estimation because the actual behaviors of the different cost drivers are not recognized. Moreover, experiences and observations show that these expenses are not really fixed if multiple

machines or operations are related to a product [Cooper and Kaplan, 1991]. The advantages of ABC can be found in the following aspects.
(1) Indirect costs are more accurately estimated in organizations producing a diverse product mix.
(2) It helps to reduce the effort to determine the relevant costs because of the improved structure of the cost model.
(3) ABC can improve decision-making and facilitate continuous improvement to reduce the overhead costs. This can be achieved through a series of activity analysis, namely activity elimination, activity reduction, activity selection and activity sharing.

One limitation of ABC can be attributed to the difficulty in obtaining sufficient and accurate information to enable proper cost allocation.

Table 6.5 Classification of activities

Level		Definition	Content
High	Facility-sustaining	Activities for operating the facility for the period, which are not associated with specific products.	Administration (*e.g.* executive salary), taxes, housekeeping, landscaping, maintenance, etc.
↕	Product-sustaining	Activities that are associated with the individual products but are independent of how many units or batches of the product are produced.	Product specifications, engineering change notice, product enhancement, information system.
	Batch-level	Activities that are related to each batch of goods processed. The resources consumed are independent of the number of units produced.	Machine setting-up or teardown, purchase orders, inspection, material handling.
Low	Unit-level	Activities related directly to the production rate.	Direct labor, material, machine costs, energy.

6.2.2.8 Design to cost

As a converse measure to traditional cost estimation and control, design to cost (DTC) strives to make design converges on cost instead of allowing cost to converge on design [Michaels and Wood, 1989]. DTC mandates cost success through the making of products or systems affordable to the customers. To do so, cost-realistic products are set in the early design stage, along with performance and schedule. Activities and responsibilities are clearly defined for all departments. In addition, development schedule is strictly controlled. Hence, the resulting products are inherently lower in cost.

Based on the above discussions, Table 6.6 summarizes the characteristics of the models.

Table 6.6 Comparison of cost models

Cost model	Cost driver	Applicability	Accuracy
Expert opinion	Expert judgment	Product planning stage	Low
Analogy	Project similarity	Conceptual stage	Fair
Function costing	Product function	Conceptual stage	Fair
Parametric costing	Design parameters, such as weight or number of parts, unit power cost, etc.	Embodiment & detailed design stage	Middle
Productive hour costing	Productive hour rate cost; Indirect costs	Detailed design, process planning	Middle
Magnitude-based modeling	Component cost level	Embodiment & detailed design stage	Fair
Activity-based costing	Activity	Embodiment, detailed design, process planning	High

6.3 Cost Estimation in Product Family Development

Section 6.2 presents the fundamental issues of and the methods for cost estimation for product development. In this section, the scope is narrowed down to cost estimation of a product family. As mentioned earlier, product family design follows the paradigm of mass customization, which aims at increasing the product variety while maintaining mass production efficiency [Pine, 1993a]. It represents an effort to achieve the economy of scale and the economy of scope simultaneously. Thus, cost is an integral part of product family design. Efficient cost estimation of product family is required.

Cost can be estimated using various measurements, such as time, commonality and monetary measures. An estimation of commonality can be traced to a sizable body of research in management science and operation research [Collier, 1981; Treleven and Wacker, 1987; Martin and Ishii, 1996, 1997; McAdams et al., 1999; Jiao and Tseng, 2000]. A commonality index is constantly used as the measurement of commonality of a product family (Section 6.3.1). Next, cost estimation based on monetary measurements are studied, which provide a more accurate estimation based on an empirical analysis of the cost elements and their relations (Section 6.3.2).

6.3.1 *Commonality index*

The commonality index is a measurement of the internal similarities between product variants within a product family. Collier [1981] proposed the degree of commonality index (DCI) as a metrics of the commonality underlying a product architecture based on the company's BOM. DCI can be used to measure commonality at different levels, namely a single product, a product family or a product line. Wacker and Treleven [1986] extended the DCI and developed the total constant commonality index (TCCI), which distinguishes commonalities within a product from those between products. Furthermore, Treleven and Wacker [1987] explored the process commonality based on set-up time, flexibility in sequencing and flexibility expediting decisions. Jiao and Tseng [2000] developed a commonality index that incorporates a component commonality and the process commonality into a unified formulation. Kota et al. [2000] established a product line commonality index to assess the commonality levels of a product family based on various manufacturing factors, such as size, shape, material, processes, assembly, etc. Siddique [2000] proposed two measures, namely a component commonality and a connection commonality, and applied them to the modularity analysis of automobile underbodies. Martin and Ishii [1996, 1997] used three indices, namely a commonality index, a differentiation index, and a setup index, to cater to the cost of product variety. McAdams et al. [1999] studied the similarity measure of a product family from a functional perspective and proposed the functional similarity index to assist modular product design.

Commonality is closely related to production cost, under the assumption that maximizing commonality among products minimizes the production cost [Simpson, 2004]. Thus, a sensible commonality can help estimate and improve cost-efficiency in product family design. However, the commonality index is a vague term and is not intuitive. For more accurate cost estimations, it is preferable to model the market demands and the associated manufacturing costs directly using monetary measurements.

6.3.1.1 Cost estimation in monetary measurement

Monetary measurements allow for a more accurate estimation and a better control of the solution optimality. The cost models thus developed depend on the analysis of various cost elements, such as materials, machine time, direct labor, overhead, etc. The cost models proposed in Section 6.2, such as the function costing, MBC, ABC, etc., are also applicable to product family cost estimation, given that a cost element can be logically identified and allocated.

In product family design, cost estimation is usually presented as a combination of these modeling techniques. Chidambaram and Agogino [1999] proposed a model that combines the generational cost, which is based on the fundamental elements and features of a product, and the function-based cost. Regression-based and similarity-based methods are used to compute the cost. Fujita *et al.* [1999] developed a cost model by considering the fixed cost and the variable cost. Learning effect is considered when computing the fabrication cost and the assembly cost, which are components of the variable cost. However, these methods require a lot of downstream manufacturing information, which may not be available at the product planning stage. Thus, a number of coefficients and parameters have been assigned subjectively. Gonzalez-Zugasti *et al.* [2001] proposed a cost model that accounts for the investment on platforms and product variants. An optimization process is carried out to search for the optimal solutions. Uncertainties during the development of a product family are also considered and addressed using the concept of real options. Van Wie *et al.* [2001] developed guidelines for redesign to reduce assembly cost based on an analysis of the module interface complexity. Grante and Andersson [2003] considered three factors in the cost model, namely, the part cost, development cost and production cost. Park and Simpson [2005] proposed a comprehensive cost model based on ABC. However, this model is restrictive in that production activities and resource data are usually not known to the designers at the design stage.

These are representative cost estimation applications in product family design. A common deficiency of these cost models is their reliance on the detailed knowledge of the product design and the related process plan, intertwined with the uncertainty of such knowledge at the early design stage [Jiao and Tseng, 2004]. These have made cost estimation a formidable task. To overcome this difficulty, cost estimation based on prior knowledge and careful decomposition of cost elements is necessary.

6.4 An Empirical Cost Model for Design Reuse

In this book, a cost model is developed as part of the design reuse methodology. This cost model accommodates the requirements of cost estimation of product family design based on design reuse. In particular, the following characteristics are identified in reuse-based product family design.

(1) The design reuse methodology assumes a modular product architecture. Hence, basic physical modules can be identified as the costing elements.
(2) The cost of modules accounts for a large portion of the production cost.
(3) Based on the design reuse rationale, the cost of the modules can be estimated and predicted based on historical data, with considerable reliability.
(4) Cost estimation is carried out at the product planning and conceptual design stage. Detailed information of manufacturing resources is not available and hence, not included in the cost model.

170 *Design Reuse in Product Development Modeling, Analysis and Optimization*

The cost model considers three major elements, namely, the fixed cost, development cost and component cost. These elements are defined at different levels of the product family development (Figure 6.4). Accordingly, the cost model in mathematical form is presented in Equations (6.3) and (6.4).

$$C = C_F + \sum_{i=1}^{N} C_P^i \qquad (6.3)$$

$$C_P^i = C_d^i + \sum_{j=1}^{M^i} C_c^j \qquad (6.4)$$

where C_F is the fixed cost at the corporation or department level; C_P^i is the cost of product i; N is the total number of products in the product family; C_d^i is the development cost of product i; C_c^j is the cost of component j; and M^i is the number of components implemented in product i.

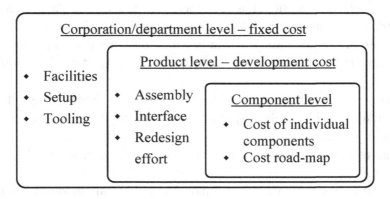

Figure 6.4 Cost model of a product family

6.4.1 *Fixed cost*

The fixed cost is estimated at the corporation or department level. It includes the costs required to maintain the regular operations of a manufacturing unit, such as the cost incurred for the maintenance of the manufacturing facilities, the setup and tooling costs, etc. This part of the cost is not sensitive to the changes in product configurations, and remains constant over a period of time. The fixed cost is estimated based on the historical manufacturing data. It should be noted that, in this book, the fixed cost refers to the fixed cost of a product family. If several product families are being designed simultaneously, it is necessary to apportion the fixed cost among them.

6.4.2 Development cost

The development cost is applied at the product level. Apart from the cost of individual components, cost is incurred to assemble the components, or to analyze and test the performance of the products. Factors to be considered include the characteristics of the interfaces, the complexity of the assembly, and the fraction of a design that is new. Different criteria can be applied to estimate the development cost, such as the development time [Grante and Andersson, 2003], number of primitives [Fujita *et al.*, 1999], product weight [Fujita *et al.*, 1998a], etc.

This book adopts a model that accounts for the complexity caused by the number of components (N_c), the number of types of components (N_t), and the number of interfaces (N_i). First, the complexity of a product is estimated based on these factors. Next, this complexity is used to approximate the development cost. The development cost of product i is calculated using Equation (6.5) [Meyer and Lehnerd, 1997].

$$C_d^i = \alpha^i \sqrt[3]{N_c N_t N_i} \qquad (6.5)$$

where α^i is the cost coefficient of product complexity, and $\sqrt[3]{N_c N_t N_i}$ is used to measure the product complexity.

6.4.3 Component cost

The component cost is the most important part of the cost model. It accounts for the majority of the product cost. Currently, production based on an 'assemble-to-order' practice is prevalent in modular design. Product variants can be generated based on the physical components, which can be bought 'off the shelf'. Thus, the component cost can be used as a basic costing element. In this way, the lower level product cost information, such as the primitive manufacturing elements (*e.g.*, material, machining, labor, etc.), is not required. By analyzing the historical data, cost road-maps can be built for the components to indicate the trend of the cost evolution.

The production volume is an important factor that influences the unit cost of a component. A higher production volume usually leads to a lower unit cost because of the possible quantity discounts. Moreover, cost usually decreases with time due to the learning effect and technology advancement. As an example, the cost road-maps of the two cellular phone batteries are illustrated in Figure 6.5. Each curve represents a road-map of the unit cost of a component at a particular production volume per year (the number shown on top of the curve).

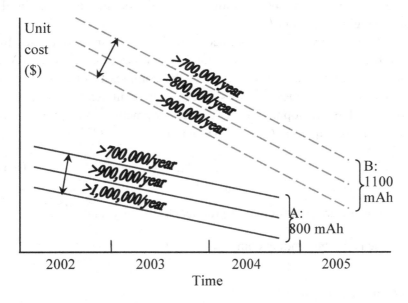

Figure 6.5 Cost road-maps of cellular phone batteries

This section presents a cost model for product family design. Analysis of existing products and manufacturing systems is emphasized in this model, which makes it suitable to the design reuse methodology. The cost elements can be refined with respect to a specific product family to be designed. The overall cost of a product family is estimated based on the model once the configurations of the products are generated.

6.5 Summary

Cost is an important criterion to evaluate product optimality. This chapter presented an effort to investigate cost estimation methods. The cost structures and cost modeling techniques are the two interrelated issues in cost estimation. The former defines the basic cost elements/drivers that are necessary in cost estimation. The latter is the aggregation scheme by which the cost elements are combined into a logical formulation. Based on the bottom-up and top-down perspectives of the cost structure, a number of cost models are investigated and their pros and cons are discussed.

Cost estimation for product family design is discussed according to two strategies.
(1) The commonality index is a measure of the commonality within a product and among product variants. It provides useful guidelines to reduce the product cost.
(2) The cost estimation using monetary measure provides a more straightforward and accurate estimation.

As a useful attempt, a cost model for the evaluation of a product family cost based on historical product data is proposed. This model involves cost elements at different levels of production. The fixed cost is defined at the department level; the product development cost is defined at the product level; and the component cost is defined at the component level. A cost-road map is used to model the evolution of costs at the component level, which is suitable to estimate component cost based on the design reuse methodology.

Chapter 7

Product Performance Evaluation

This chapter focuses on the evaluation of product performance. Performance evaluation requires a definition of the evaluation criteria and the subsequent evaluation based on these criteria (Section 7.1). Product performance is closely related to the concept of product quality and robust design, which is discussed in Section 7.2. In Section 7.3, the information content assessment (ICA) method is proposed to evaluate product performance based on the principles of axiomatic design [Suh, 2001]. This method involves systematic steps to identify the component capability index, and the calculation of the product information content. A few precautions are discussed for the application of this method. Finally, a case study is presented to showcase the ICA method (Section 7.4).

7.1 Introduction

To effectively carry out performance evaluation, two issues are critical, namely, (1) the establishment of the relationship between the performance and the design parameters (DPs), and (2) the formulation of a common basis to accommodate the diverse performance criteria that are inherently incommensurable.

7.1.1 *Relating performance to design parameters*

DPs constitute the internal elements of a product that are intended to achieve a specific performance. Given the external signal, product performance is largely determined by the DPs, and hence the design process involves determining the right values for the DPs to achieve the desired performance (Figure 7.1). Thus, a relationship between performance and the DPs is necessary if a designer wishes to make the right decisions. A straightforward method is to establish equations between the DPs and the output performance. Given the external signal i, the performance y, and the DPs x, the relationship can be illustrated as in Figure 7.1.

$$y = f(x, i)$$

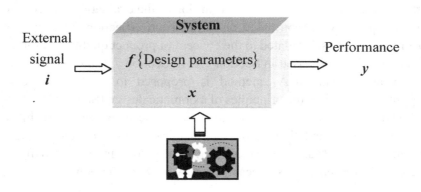

Figure 7.1 Relationship between performance and design parameters (DPs)

When such equations can be established, the estimation can be accurate. Using the example of Equation (3.1) in Chapter 3, the behavior of a heating device can be expressed as $E=I^2R$, which is the relationship between the performance (energy E) and the parameter (resistance R), with respect to the external signal (current I).

However, in a typical engineering problem, it is often difficult, if not impossible, to formulate appropriate equations. Numerical methods, such as finite element analysis (FEA) and computational fluid dynamics (CFD), can be used to overcome this difficulty. However, the limited power of computers and the complexity of the programming codes restrict the application of these numerical methods. The presence of uncertainties, variations and noise factors would worsen the problem.

Alternatively, meta-models can be used to simulate the equations. A meta-model is a 'model of the model' [Kleijnen, 1987], and it is used when the true nature of the relationship between the performance (y) and the DPs (x) cannot be captured precisely. A meta-model can be established using statistical techniques such that

$$\tilde{y} = g(x,i),$$

and hence,

$$y = \tilde{y} + e,$$

where e represents an approximation or measurement error.

A meta-model not only yields insight into the relationship between the output performance and the input DPs, but also provides a rapid analysis tool that is inexpensive to implement. The meta-modeling techniques that have been used in product design include design of experiments (DOE) [Simpson, 1998; Ulrich and Eppinger, 2004], response surface method (RSM) [Chen *et al.*, 1996; Simpson *et al.*, 1996; Jiao and Tseng, 2004], kriging [Simpson, 1998], regression analysis [Fujita and Yoshioka, 2003], etc.

A pre-requisite of meta-modeling is that experimental data can be obtained from existing cases or from simulation results. However, it is not always possible to satisfy this condition. If invalid experimental data is used to build the models, these models can be invalid.

7.1.2 *Aggregating performance criteria*

For the second issue, it is desirable to aggregate the diverse measures of design performance into a unified metric. A possible solution is to normalize the multiple performance criteria and aggregate them into a utility function, such as the capability index [Simpson *et al.*, 1997a; Harry and Schroeder, 2000], deviation function [Simpson *et al.*, 2001; Hernandez *et al.*, 2002], and product portfolio utility [Mangun and Thurston, 2002]. However, the inherent incompatibility between different criteria and the arbitrary assignment of weights to these criteria hinder the effectiveness of such utility functions.

Recently, there is an advocate to use information content as a unified measurement of multiple criteria. Information content is defined in terms of the probability that the design requirements can be satisfied by the DPs [Suh, 2001]. It is a dimensionless metric that can be computed using statistical techniques. The major concern is the formulation of the problem and the computation of the information content.

Information content was calculated with the support of fuzzy QFD, and used to estimate the critical characteristics of routine design [Bahrami, 1994]. The limitation of this method is that the QFD process could only roughly estimate the customers' preferences. A concept evaluation method based on information content and fuzzy ranking was used in configuration design for mass customization [Jiao and Tseng, 1998]. An appealing feature of this method is that it combines 'tangible' and 'intangible' criteria into the same evaluation metric. Information content was used to measure design customizability in mass customization, where the system ranges and design ranges, which are necessary to compute the information content, were assumed to be known [Jiao and Tseng, 2004]. However, it is not easy to obtain the system ranges and the design ranges. In actual fact, information content as an evaluation metric has not been widely adopted, mainly due to the difficulties to formulate the design ranges and the system ranges. Therefore, proper procedures to derive these two ranges have to be developed in order to apply this approach.

7.2 Robust Design

Product performance evaluation and control is an important issue in quality engineering and robust design, which host a substantial body of research and applications since 1980s when the seminal Taguchi method was introduced in the United States [Clausing, 1994]. The Taguchi method provides solutions to incorporate product quality into the design processes. This section provides an overview of robust design based on the Taguchi method. The discussion is helpful to the readers to understand the performance issues presented in this Chapter.

Firstly, it is important to define robust design, and what constitutes product quality. Quality can be defined from two aspects, namely, (1) the features of a product, and (2) the conformance of this product to these features [Juran and Gryna, 1988]. The features of a product are the performance characteristics that attract the potential customers. Conformance to the features refers to the ability of the product to consistently deliver the desired performance throughout the PLC, under all intended operating conditions. Such a definition also applies to the quality of a process that is used to produce the relevant product.

The robustness of a product is related closely to the second aspect of product quality, *i.e.*, conformance to the intended features. Specifically, the quality of a product can be influenced by many factors, as shown in the P-Diagram in Figure 7.2. A P-Diagram is essentially a schematic diagram that relates the input (signal factor) to the desired output (response). The signal factor is transformed via the control factors, which are typically elements that are under the 'control' of the designer, such as design, materials and processes. The noise factors are factors that can influence the design but are not under the control of the designer, such as the environmental factors, deterioration over time, unit-to-unit variation, etc. The noise factors could cause the performance characteristics of a product to deviate from the target value. Robust design aims to alleviate or eliminate the effect of the noise factors. There are two approaches to achieve this, namely, (1) eliminating the source of the noise, and (2) eliminating the sensitivity of a product to the noise. The former is usually very difficult and costly to be achieved. Hence, robust design often focuses on the later approach. Accordingly, a robust product/process is one that is "insensitive to the effects of sources of variability, even through the sources themselves have not been eliminated" [Fowlkes and Creveling, 1995].

The sensitivity of a product to noise factors is illustrated in Figure 7.3. Product A and Product B are designed to achieve the same functionality. Their performance characteristics are different according to the two straight lines[1] that show the relationships between the response T, and the design parameter D_j. Both products can achieve the target performance T_0 at a specific set-point, D_0. However, considering the noise factors, D_i may vary within a certain range according to a specific distribution function. Assuming that the variations of the DP are the same for both products, the response of Product A varies in the range of $[T_0 - \Delta_A, T_0 + \Delta_A]$, and that of Product B varies in the range of $[T_0 - \Delta_B, T_0 + \Delta_B]$. It is obvious that Product A has a smaller variation and hence is less sensitive to noise factors. Product A is considered a more robust design than Product B.

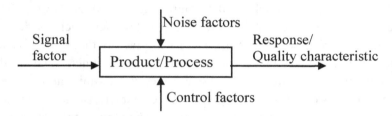

Figure 7.2 P-Diagram of a product/process

In the Taguchi method, the quality loss function is used to measure the quality of a product in terms of the total quality loss of the product to the society [Taguchi, 1986]. The Taguchi method emphasizes off-line quality engineering, which aims to incorporate robustness/quality into a product and its manufacturing processes. It involves systematic procedures to increase the robustness of a system and reduce the quality loss. To do so, three stages are defined, namely, the system design stage, parameter design stage and the tolerance design stage [Taguchi, 1986] (Figure 7.4).

---

[1] This example uses the straight lines to denote a linear relationship between the DPs and the response. In circumstances where a non-linear relationship is applicable, the concept of robustness remains unchanged.

Product Performance Evaluation 181

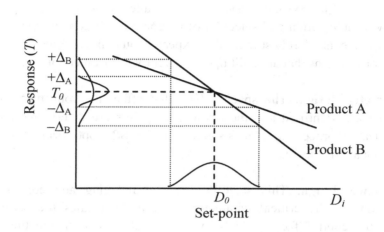

Figure 7.3 Sensitivities to design variation

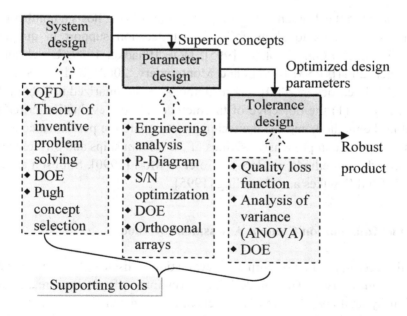

Figure 7.4 Three stages in off-line quality control

System design. System design is also called conceptual design. It involves the creation and selection of alternative solutions that embrace the ingredients of robustness. The expected outcome is one or a few superior solutions that are built upon inherently robust technology.

Parameter design. The specific DPs that affect the robustness of a product are identified and the optimal set-points are determined using optimization processes, such as signal-to-noise (S/N) optimization, DOE, orthogonal arrays, etc.

Tolerance design. This stage focuses on reducing and controlling variations in the critical parameters, *i.e.*, it determines the level of allowable control factor variability. The goal is to optimize the unit manufacturing cost, life-cycle cost and the quality loss, such that a robust design is achieved.

Other than the Taguchi method, quality engineering hosts a number of methods and tools to allow for efficient decision support in quality control, such as QFD [Cohen, 1995], DOE [Phadke, 1989; Ulrich and Eppinger, 2004], RSM [Myers and Montgomery, 2002], and Six Sigma [Harry and Schroeder, 2000]. These methods are summarized in Table 7.1 according to (1) the objective of the method, (2) the evaluation criteria used in the method, and (3) the techniques to relate the performance and DPs. For more comprehensive studies of the relationships and differences between these methods, readers can refer to Mori [1990], Simpson *et al.* [1997b] and Fowlkes and Creveling [1995].

7.3 The Information Content Assessment Method

In this section, product performance evaluation is discussed in the context of design reuse. The underlying principle of design reuse is design-by-analogy. Solutions are generated from existing components, which quality has been proven in designed products. In such a sense, the quality of a product can be better predicted and ensured. Therefore, it is reasonable to evaluate the performance of a new product based on existing similar designs.

Considering the rapid advances in technology, only elements of products belonging to the existing technology are reused in new designs. Given that a performance criterion is influenced by one or a few components (*i.e.*, the DPs), the outcome can be predicted from similar designs that make use of identical components.

Table 7.1 A summary of quality engineering approaches

	Objective	Evaluation criteria	Relation between performance & DPs
QFD	Transformation of Voice of Customer (VOC) into technical characteristics that drive the product development and process planning.	Technical priorities; Competitive benchmarking; Engineering target values.	House of Quality (HOQ); Design matrix
DOE	Understand the interactions between specific control factors and noise factors and optimize a system by eliminating the effect of noise factors.	S/N ratio; Quality loss function.	Experimental design; Inner and outer arrays; Analysis of variance (ANOVA)
RSM	Build empirical models that relate the response/output with the levels of DPs/input variables.	Least square error in model-fitting	First-order polynomial Second-order polynomial;
Six Sigma	Deliver high performance, reliability, and value to the end customer by reducing process variations that cause defects.	Capability indices	HOQ; Product planning matrix; Process planning matrix.
Taguchi method	Improve the quality of a product/process by minimizing total quality loss.	S/N ratio; Total quality loss to society.	Quality loss function

7.3.1 Background – *information axiom and information content*

Suh [2001] proposed the theory of axiomatic design in which he advocated two axioms as the basic principles for system design.

Axiom 1: *Independence Axiom*, which requires that the functional requirements be kept independent of each other.

Axiom 2: *Information Axiom*, which demands the minimum information content of a design.

Information content is defined in terms of the probability that the FRs can be satisfied by the DPs. According to the information axiom, among the solutions that satisfy the independence axiom, the solution with the minimum information content is the optimal one. In addition, information content can be used as a measurement of the complexity of a design, *i.e.*, the uncertainty in achieving the specific FRs [Suh, 2005]. Thus, given the same set of FRs, a design with a lower information content involves less complexity than the one with a higher information content, and thus is superior to the latter.

To calculate the information content with respect to a specific design requirement, first, the achievable performance must be computed from the DPs, as they are set in the designed products. The achievable performance is called the system range, which can be generalized as a probability density function (pdf). Figure 7.5 uses a normal distribution to represent the system range. At the same time, the customers set the design requirements, which can be quantified as the design ranges, usually in the form of the lower and upper bounds of the target product performance. Next, the system range is compared with the design range, resulting in a common range. Thus, a probability (ζ_i) is obtained to show how well the design requirement can be satisfied by the DPs. Information content (I_i) is computed as the logarithmic function of this probability [Suh, 2001], as shown in Equation (7.1).

$$I_i = \log_2 \frac{1}{\zeta_i} = -\log_2 \zeta_i \qquad (7.1)$$

In the case of a design task with p design requirements, the information content of the entire system (I_{sys}) is determined using Equation (7.2).

$$I_{sys} = -\log_2 \zeta_{\{p\}} \qquad (7.2)$$

where $\zeta_{\{p\}}$ is the joint probability that all requirements are satisfied.

Provided that the design requirements are independent of each other, $\zeta_{\{p\}}$ can be computed as the multiplication of the individual probabilities (ζ_i), i.e. $\zeta_{\{p\}} = \prod_{i=1}^{p} \zeta_i$. Thus, the information content of the system can be computed using Equation (7.2').

$$I_{sys} = -\log_2 \prod_{i=1}^{p} \zeta_i = -\sum_{i=1}^{p} \log_2 \zeta_i \qquad (7.2')$$

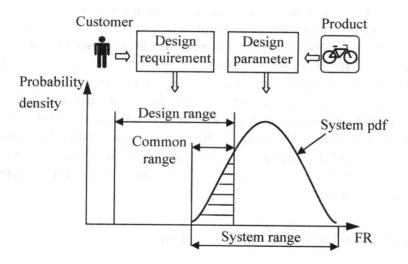

Figure 7.5 Relationship between design range and system range

Equation (7.2') indicates that for a system which design requirements are independent of each other, the information content of the system is the summation of individual information content, i.e., $I_{sys} = \sum_{i=1}^{p} I_i$. However, if the design requirements are not independent, the information content of a product cannot be computed this way. The correlations between the design requirements have to be established to compute the joint probability that the design requirements can be satisfied.

Considering the difficulties in establishing the correlations between the design requirements, it may involve tremendous efforts to compute the information content. Therefore, in this book, without the restriction of the independence of the design requirements, the summation of the individual information content is used as a measure of the overall product performance, which can be considered as the pseudo information content. For description consistency, the term 'information content' is used to refer to the pseudo information content with respect to a product, as well as the information content with respect to individual design requirements.

An attractive feature of using the information content in performance evaluation is that it is a dimensionless metric based on probability. Hence, it provides a uniform metric to incorporate various measures of technical criteria that are inherently incommensurable [Jiao and Tseng, 1998].

To effectively use the information content for performance evaluation, it is required that the system range and the design range of a product be established. However, the establishment of the system range has been difficult and lacked effective computational tools. Therefore, this book proposes the ICA method to establish the system ranges and assess the information content.

7.3.2 *The information content assessment process*

The ICA method involves six steps as shown in Figure 7.6. Each step is supported by relevant tools or information resources (shown in the ellipses). Steps 1 ~ 3 are carried out in the product information modeling and analysis stage. They are responsible for establishing the component capability indices, which define the system ranges of the physical components. Steps 4 ~ 6 are carried out at the design synthesis and evaluation stage, during which the configurations of products are generated and the information content is calculated. The steps are discussed in Sections 7.3.3 and 7.3.4, using an electric kettle product as an illustrative example.

7.3.3 *Establishing system range from existing products*

Product performance is evaluated against the KCs. Therefore, the relevant KCs have to be identified first. Moreover, the correlation matrices, TR_2 and TR_3, are required for establishing the mapping route from the KCs to the physical components in the component catalog. Steps 1 and 2 are designed to carry out these processes, supported by the KC extraction and correlation matrices, respectively. These issues have been discussed in Chapter 4 and the results for the electric kettle product are reproduced in Tables 7.2 and 7.3, respectively. In addition, Table 7.4 presents the component slots that correspond to the common product functions. The number of components within each slot is listed in the last column.

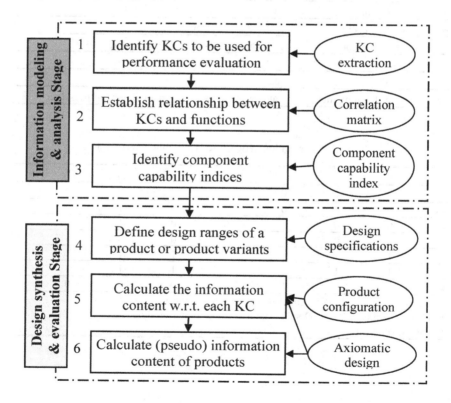

Figure 7.6 The processes of the ICA method

188 Design Reuse in Product Development Modeling, Analysis and Optimization

The key step in the ICA method is Step 3, through which the capability indices of the individual components (if a KC is related to one component) or component combinations (if a KC is related to more than one component) are extracted. The processes are illustrated in Figure 7.7 and explained using an example of a component combination of the electric kettle product.

Table 7.2 KCs of the electric kettle

KCs		Type	Unit	Default value
k_1	Power consumption	real	watt	–
k_2	Dimension	real	mm	–
k_3	Water capacity	real	liter	–
k_4	Automatic control	Boolean	–	True(1), False(0)
k_5	Cost*		S$	
k_6	MTBF	real	hour	–
k_7	Water fetching method	categorical	–	manual, tap, air pressure

Table 7.3 Correlation between KCs and functions (TR_2)

KCs	f_1	f_2	f_3	f_4	f_5	f_6	f_7
k_1	0	1	0	1	0	0	0
k_2	0	0	0	0	0	1	0
k_3	0	0	0	0	0	1	0
k_4	1	0	0	0	0	0	1
k_5	1	1	1	1	1	1	1
k_6	1	1	1	1	1	1	1
k_7	0	0	1	0	0	0	0

f_1: Keep warm f_2: Heat transfer f_3: Water fetching:
f_4: Heat generation f_5: Display status f_6: Water holding
f_7: Heating control

---

* Cost is not considered in performance evaluation because it is not a performance factor.

Table 7.4 Function and component slot

Function		Component slot		Number of components
f_1	Keep warm	s_1	Insulator	4
f_2	Heat transfer	s_2	Contact disk / radiator	3
f_3	Water fetching	s_3	Outlet device	3
f_4	Heat generation	s_4	Heating disk / coil	7
f_5	Display status	s_5	LED / LCD screen	5
f_6	Water holding	s_6	Container	6
f_7	Heating control	s_7	Thermometer and switch	3

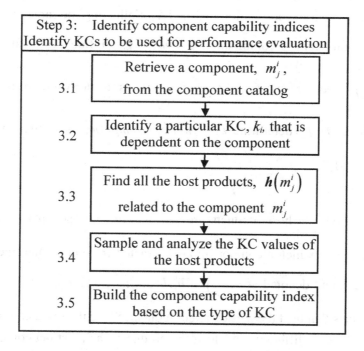

Figure 7.7 Computing the component capability indices

Step 3.1 Retrieve a component or a component combination from the component catalog, which is represented as (m_j^i), where j is the index of the component slot (s_j) and i is the index of the component in slot s_j. For example, the component combination, (m_2^1, m_4^1) = (*3-layer radiator, heating disk*) of the electric kettle product is retrieved. The capability index of (m_2^1, m_4^1) is to be established.

Step 3.2 The component is related to a particular product function f_i, through the correlation matrix TR_3, which in turn, is related to a particular KC, k_i, through the correlation matrix TR_2. Following the same example, the component combination (m_2^1, m_4^1) influences functions (f_2, f_4) = (*heat transfer, heat generation*), as is derivable from TR_3. Functions (f_2, f_4) determine k_1=*power consumption*, as shown by TR_2 (Table 7.3).

Step 3.3 All the host products are retrieved for component (m_j^i). The product cases which a component are extracted from are called the **host products**, and denoted as $h(m_j^i) = \left[h_j^1, h_j^2, ..., h_j^u \right]$, where u is the number of products that host the particular component. Since the same component may be used in different models of the products, each distinct component or component combination may be related to a number of host products. In this case study, the component combination (m_2^1, m_4^1) is used in three host products, namely $h(m_2^1, m_4^1) = \left[h^1, h^2, h^3 \right]$ as shown in Table 7.5.

Step 3.4 With respect to k_i, the KC value of each host product is obtained. The KC values can be discrete or continuous depending on the type of KC. For the continuous type of KC, the KC values usually vary in a certain range. Several samples can be retrieved for each host product, and their KC values are obtained. For example, with respect to k_1: *power consumption*, ten samples are retrieved for every product h^i ($i=1, 2, 3$) (Table 7.5). The *power consumption* value of these samples can be identified based on historical data.

Step 3.5 The KC values related to the sampled host products are merged to form a distribution function, known as the capability index of a component. This can be done manually if the distribution is simple. Alternatively, for a more complex data pattern, the distribution function can be built using statistical methods, such as the parametric estimation [Siddall, 1983]. Thus, the capability index can be established, in the form of a pdf for a continuous type of KC, or a probability mass function (pmf) for a discrete type of KC. Using the same example, the pdf for the component combination $\left(m_2^1, m_4^1\right)$ can be formulated as a normal distribution based on the sample data in Table 7.5. The capability index is defined as $\left\{\left(m_2^1, m_4^1\right) \mid k_1 : \mu = 1499, \sigma = 20.6\right\}$ and is plotted in Figure 7.8.

A few typical distribution functions for representing the capability indices with respect to different types of KCs are shown in Figure 7.9.

In general, the capability indices of the components can be established using these steps. However, there is one situation where these steps are not effective. This situation arises when a KC is related to a number of functions, which in turn are related to a number of physical components, such that there could be too few component combinations available from existing product cases. In such a situation, it is preferable to identify the exact relationships between the attributes of the components and the KCs. Such relationships can be in the form of equations or meta-models. Accordingly, the system range can be computed as follows.

Table 7.5 Sampled *power consumption* values of three products that host $\left(m_2^1, m_4^1\right)$

Sample	h^1	h^2	h^3
1	1461	1490	1517
2	1475	1505	1516
3	1471	1484	1506
4	1475	1514	1509
5	1474	1519	1512
6	1477	1514	1534
7	1512	1531	1540
8	1480	1516	1528
9	1447	1499	1498
10	1496	1498	1499

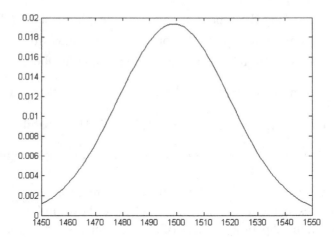

Figure 7.8 Capability index for component combination $\left(m_2^1, m_4^1\right)$

Assuming that the product performance is denoted as *k*, and the related component attributes are denoted as a vector *ı*=[$ı_1$, $ı_2$,..., $ı_q$], an equation can be established as *k*=*ψ*(*ı*), where *ψ* is the mapping function. Given that the distribution of the parameter *ı* is known, the distribution of the product performance *k* can be estimated using the first order Taylor series expansion of the function [Nayak *et al.*, 2002]. For example, if the mean and variance of the parameter ι are μ_{ι_i} and $\sigma_{\iota_i}^2$ ($i=1, 2,..., q$), the mean and variance of the performance k can be computed as:

$$\mu_k = \psi\left(\mu_{\iota_1}, \mu_{\iota_2}, ..., \mu_{\iota_q}\right)$$

$$\sigma_k^2 = \sum_{i=1}^{q} \left(\frac{\partial \psi}{\partial \iota_i}\right)^2 \sigma_{\iota_i}^2$$

Thus, the distribution function that defines the system range can be established. This is an alternative to the correlation matrix-based method to establish the capability indices.

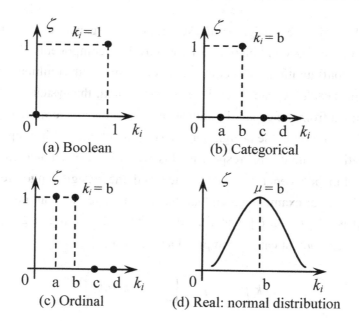

Figure 7.9 Typical capability indices for different types of KCs

7.3.4 *Assessing information content*

Typically in product family design, a ranged set of design requirements is initiated to cover the target market segmentations. This corresponds to

Step 4 of the ICA method. The design requirements are interpreted as a set of design ranges according to the KCs. For example, the designer may wish to launch a new model of electric kettle P_1, which design requirements are defined in Table 7.6.

Table 7.6 Design requirements of a family of electric kettle products

KCs	P_1
Power consumption (W)	$800 \pm 5\%$
Water capacity (L)	1.7~1.8
Automatic control	True
MTBF (h)	$\geq 20,000$
Water fetching method	manual

In Step 5, the design synthesis algorithms generate the product configurations as combinations of the physical components. From the product configurations, the component or component combination that determines each KC can be easily identified. Next, the system range can be obtained from the capability indices that have been determined in Step 3. At the same time, the design range has been determined in Step 4. The information content with respect to this KC can be calculated from the relationship between the system range and the design range (refer to Figure 7.5). For example, given that the design range of a continuous type KC (k_i) is [dn_L, dn_U], and the pdf of the system range is $\zeta(k_i)$, the information content can be computed as Equation (7.3).

$$I(k_i) = -\log_2 \left(\int_{dn\_L}^{dn\_U} \zeta(k_i) dk_i \right) \qquad (7.3)$$

Equation (7.3) is a special case of Equation (7.1), *i.e.*, when a particular KC has a continuous value and the system range, pdf, is known. For the electric kettle example, the configuration for P_1 is presented in Table 7.7. Accordingly, the information content with respect to the KCs is computed in Table 7.8.

Table 7.7 Product configurations of an electric kettle

Component slot		Component used in P_1	
s_1	Insulation	m_1	nil
s_2	Contact disk / radiator	m_2	Copper radiator
s_3	Outlet device	m_3	Tap outlet
s_4	Heating disk / coil	m_4	Resistance coil 3.5A
s_5	LED / screen	m_5	Boiling LED
s_6	Container	m_6	1.75L Plastic container
s_7	Thermometer and switch	m_7	Auto power cut

Finally, in Step 6, the pseudo information content of a product is computed. Based on Equation (7.2'), the pseudo information content of electric kettle P_1 is:

$$I'(P_1) = \sum_{i=1}^{5} I(k_i) = 0.11 + 0 + 0 + 0.03 + 0 = 0.14.$$

Information content is a measurement of the probability that the design requirements can be satisfied based on the available capability of the designed cases. Using the reverse form of Equation (7.1), the probability that the design requirements can be satisfied can be computed from the information content:

$$\zeta_i = 2^{-I_i} \tag{7.4}$$

The information content of the electric kettle product P_1 is 0.14, which means that the probability that all the design requirements can be satisfied was approximately 90%. This gives a designer an idea of the overall performance of the product. The designer can focus on one or a few KCs, if the information content related to them is significantly larger than the others. The components related to those KCs can thus be modified to enhance the relevant product performance.

Information content integrates various product performance criteria into a scalar value, which makes it convenient to compare different products. In addition, the design synthesis problem can be formulated into an optimization problem that allows for easy computational support.

Table 7.8 Computation of information content

KC	Design range	Component	System range	Probability	Information content
Power consumption	[760, 840]	(m_2, m_4)	μ=794 σ=21.5	ζ=0.9269	0.11
Water capacity	[1.7, 1.8]	(m_6)	μ=1.75 σ=0.015	ζ=0.9991	0
Automatic control	True	(m_1, m_7)	True	ζ=1	0
MTBF	≥20,000	(m_4)	μ=22,500 σ=1,255	ζ=0.9768	0.03
Water fetching method	manual	(m_3)	manual	ζ=1	0

These six steps define the systematic procedures to compute the information content. The system range is extracted from the data of the existing product cases, and is generated using statistical tools. This is a useful improvement to previous research where the system range is assumed to be known [Bahrami, 1994; Jiao and Tseng, 1998, 2004], or constructed based on extensive computation and detailed analysis, which is representative of the current studies.

7.3.5 *Precautions and limitations*

As can be seen from the ICA process, the performance of a new product is determined from the performance of existing similar products based on the principle of design-by-analogy. The relationships between the performance and the DPs are established using the correlation matrices. In comparison to equation-based or meta-model-based approaches, the correlation-matrix-based relationship is not straightforward and conceals some design details. The book advocates this method for the following reasons.

(1) At the early design stage, it is very difficult to build the equations or the meta-models that are required to carry out the evaluations, mainly due to the paucity of the information. Even if such relations can be established, the uncertainties and variations in design would make their application questionable [Ulrich and Eppinger, 2004]. Therefore, it is advisable to make the estimations according to the availability of the information.

(2) A major concern of performance evaluation is the ultimate performance outcome instead of the details of the component attributes. In comparison with the detailed *content* of the DPs, the *consequence* is of greater importance [Jiao and Tseng, 2004]. This is the essence of design reuse and design-by-analogy, in which the need for detailed information of the component attributes and their relationships can be avoided.

In Step 5 of the ICA method, one situation has been overlooked, namely, the design synthesis may generate component combinations which capability indices have not been defined in Step 3. This is because the capability indices have been established according to existing product cases, and all the possible component combinations may not have been used in these product cases. In other words, there may be no host product for a particular component combination. In such a situation, the newly generated component combination (denoted as A) is compared with all the component combinations, which capability indices have already been defined. The combination that is most similar to A is selected. Similarity is measured based on the major attributes of a component, *i.e.*, the parameters that determine the characteristics of a component. The capability index of the selected similar component combination is used as the capability index of A.

To compute the information content of a product, it is preferable that the KCs are independent of each other. According to the rationale of design reuse, the KCs denote the most important factors that signify the performance of a product family. Hence, each KC represents a unique feature of the product, such that the interdependency between KCs is insignificant. However, strict independence between KCs may be difficult to be achieved, especially for variant design. In addition, the KCs in this book are not equivalent to the FRs in axiomatic design, and the design method was developed without the restriction of uncoupled or decoupled design. Therefore, the summation of individual information content is an approximation of the information content of the product. However, the purpose of the ICA method is to develop a measurement of the product performance, such that it can guide the optimization of a product family. As long as it serves this purpose, such an approximation is acceptable.

7.3.6 *A case study*

The ICA method is used to evaluate product performance of the FFU, following the case study presented in Chapter 5. Previously, the FFU was designed using the automated design synthesis approach based on the MOSGA. Product performance was ensured by optimizing the design objectives, subject to a set of constraints. In the new case study, the design problem was reformulated into a more concise form (Figure 7.10). In particular, information content was used as an integrated measure of the product performance to replace the set of design constraints that limit the design space. Accordingly, the design objectives were reduced to minimizing product cost and minimizing information content, subject to the compatibility relations between the design components. This formulation is much easier to implement. It also provides a means to measure and predict the performance levels of the designed product.

The design requirements are reproduced from the design constraints used in previous case study, as shown in Table 7.9. A new session of design synthesis based on the new problem formulation is carried out using the MOSGA method. The simplified configuration of the new product is shown in Table 7.10. The subsequent discussions focus on the performance evaluation of the new product.

Min. $T(m) = [I(m), C(m)]$
s.t. $m \in |S|$
where:
  $I(m)$ is the average information content of the product (family).
  $C(m)$ is the cost of product (family).
  m is a vector containing the physical components. Its elements are considered as the design parameters, which constitute the parameter space S.
  $|S|$ is the feasible design space, $|S| \subset S$. Feasibility is defined with regard to the compatibility among the physical components.

Figure 7.10 New problem formulation of design synthesis and evaluation

Information content was used to evaluate product performance. The details to compute the information content of the solution is presented in Table 7.11.

As can be seen from the individual information content values, the information content with respect to the six design requirements (*MT, SC, NL, CS, CM, MO*) were 0. According to Equation (7.4), it means that the probability that these design requirements can be safely satisfied is 1.

Table 7.9 Design requirements of the FFU product

KCs	Value
motor type	AC-1PH
air quantity	$1375 m^3/h \pm 10\%$
air velocity	$0.42 m/s \pm 10\%$
air uniformity	$\leq 15\%$
service cleanliness	Class1000
noise level	$\leq 50 dBA$
casing material	aluminum
casing size	3×4
vibration	$<0.025G$
mounting type	Horizontal

Table 7.10 The new product configuration

Slot	Component
Motor	R4E 355-AL02-01
Controller	SLC-S1
Filter	HEPA 70+
Casing	AL shaft-standard-A3-4
Inlet cone	Circular 1
Air guiders	3-set with array holes
Insulation	Rock-wool and fiber glass

Three design requirements (w.r.t. AQ, AV and AU) can be satisfied with high probability. In particular, the probabilities of meeting these performance targets are all above 99%. These can be derived from the information content values: 0.003, 0.007, and 0.001.

Table 7.11 Computation of information content

KC	Design range	System range	Information content
motor type (MT)	AC-1PH	$\zeta_1 = \begin{cases} 1, & MT\text{=AC-1PH}; \\ 0, & MT\text{=AC-3PH|EC}. \end{cases}$	0.000
air quantity (AQ)	[1237.5, 1512.5]	Normal distribution: $\mu=1449$, $\sigma=22$	0.003
air velocity (AV)	[0.378, 0.462]	Normal distribution: $\mu=0.42$, $\sigma=0.015$	0.007
air uniformity (AU)	≤15%	Normal distribution: $\mu=0.11$, $\sigma=0.012$	0.001
service cleanliness (SC)	1000	$\zeta_5 = \begin{cases} 1, SC\text{=1000|10000}; \\ 0, & SC\text{=1|10|100}. \end{cases}$	0.000
noise level (NL)	50	$\zeta_6 = \begin{cases} 0, & NL \leq 47; \\ 0.5(NL-47), & 47 < NL \leq 49; \\ 1, & NL > 49. \end{cases}$	0.000
casing size (CS)	3×4	$\zeta_7 = \begin{cases} 1, & CS\text{=3}\times\text{4}; \\ 0, CS\text{=2}\times\text{4|4}\times\text{4}. \end{cases}$	0.000
casing material (CM)	aluminum	$\zeta_8 = \begin{cases} 1, CM\text{=Aluminum} \\ 0, & \text{others}. \end{cases}$	0.000
vibration (VB)	≤0.025	Normal distribution: $\mu=0.0235$, $\sigma=0.0008$	0.044
mounting type (MO)	Horizontal	$\zeta_{10} = \begin{cases} 1, MO\text{=Horizontal}; \\ 0, & MO\text{=Vertical}. \end{cases}$	0.000

The largest information content was detected in the KC VB, 0.044, indicating a 97% probability that the vibration can be kept below the desired level. Since the vibration feature is mainly dependent on the attributes of the motor, which can be adjusted slightly to change the vibration level, this KC can be satisfied without major changes in product configuration.

The pseudo product information content is 0.055, which is the summation of the information content with respect to the individual KCs. From this, it is estimated that the probability that all design requirements can be satisfied is 96%.

From the analysis of the individual information content, it is expected that the product performance can be achieved by reusing the components.

To validate the evaluation of product performance, a prototype product was produced and the performance was tested. The results are presented in Table 7.12. As can be seen from Table 7.12, all performance requirements have been satisfied. The performance of AQ, AV, and AU is slightly different from the design targets, which agreed with the estimation of the information content. The design requirement with respect to VB is met by adjusting the motor attributes. These results show that information content is effective in the evaluation of the product performance.

Table 7.12 Performance of the new product based on testing

KC	Value
motor type	AC-1PH
air quantity	$1385 m^3/h$
air velocity	$0.425 m/s$
air uniformity	13%
service cleanliness	1000
noise level	$49.2 dBA$
casing material	aluminium
casing size	3×4
vibration	$0.025 G$
mounting type	Horizontal

The following remarks can be made for this case study.
(1) The characteristics of the new product can be better predicted. This is because the product configuration is generated based on existing components. The ICA method provides systematic ways to estimate the product characteristics based on existing components.
(2) The ICA method was used to evaluate product performance. It proposes a way to formulate the capability of the product components, and uses it to evaluate the product performance. As long as the initial product information is collected and formulated in the design reuse system, new products can be easily evaluated.
(3) The design reuse method can provide useful guidelines for product design, and help to reduce the need for redesign.

7.4 Summary

This chapter caters to the product performance and quality issues. Two important issues are emphasized, namely, (1) the relationship between the performance and the DPs, and (2) the metrics to aggregate multiple performance criteria. Performance evaluation is closely related to quality engineering. The Taguchi's robust design method is discussed to provide a useful insight to product quality issues. In addition, a few prevailing methods in quality control are studied according to their relation to these two issues.

An ICA method to evaluate product performance based on information content is proposed. The ICA method has its roots in Suh's axiomatic design, and it incorporates the ingredients of design reuse. Two critical issues are solved in this method, namely, (1) the formation of system ranges of existing product components in the form of component capability index, and (2) the systematic procedure to predict product performance based on the information content. From a case study, it is demonstrated that the ICA method can be used to deal with the performance estimation with reasonable applicability and credibility.

Chapter 8

A Product Family Design Reuse Methodology

The major processes and enabling technologies in design reuse have been discussed in the previous chapters. These technologies can be applied in product family design for smarter and quicker decision support. However, these technologies are not necessarily supportive by themselves. An integrated decision framework has to be developed to integrate these technologies such that they can contribute to the design from different aspects. This chapter introduces an integrated product family design reuse methodology. A prototype product design system was proposed to integrate the design reuse tools. Two comprehensive case studies are presented to show how the system works and its effectiveness in comparison with other approaches.

8.1 Introduction

In this information age, engineering design is becoming increasingly information-intensive. Decision-making relies heavily on the relevant knowledge gained from thorough analyses, modeling and simulation, previous designs and innovative concepts. Design reuse is a methodology to amalgamate the methods and tools in modeling, analysis and optimization for engineering design based on existing products or technologies. Thus, design reuse renders itself as a powerful enabler to overcome a major obstacle in product family design, namely information deficiency and uncertainty. A number of techniques under the design reuse umbrella have been introduced in previous chapters. However, these methods are limited in decision support when applied individually. To achieve effective decision support in product family design, it is necessary to combine these technologies into an integrated decision framework.

An integrated decision framework must support the multiple stages of product family design. In particular, it has to accommodate four major issues, namely *information modeling, information analysis, design synthesis* and *solution evaluation*. Based on previous chapters, the product family design approaches and their support to the four major issues are summarized in Table 8.1. These approaches are roughly divided into five categories, namely scale-based, model-based, graph-based, module-based and others (which include a few miscellaneous approaches). Each method is examined with respect to its support to the four major issues. Accordingly, a product family design reuse (PFDR) methodology is proposed in the next section to address these issues.

8.1.1 *Scale-based approach*

The scale-based approaches are based on the top-down design rationale, which does not emphasize the utilization of existing product information. Therefore, information modeling is absent in these approaches. The building of the product architecture and the design synthesis and evaluation have been formulated into a combined process, typically supported by computational tools. Multi-objective optimization has been used in performance evaluation, and the performance has been correlated with the design variables using equations or meta-models, such as design of experiments (DOE) and response surface method (RSM). On the other hand, configuration design is not relevant in these methods. Cost has rarely been considered as an optimization objective, which may be due to the lack of information to make up-front estimations. The lack of information is characteristic of the top-down approaches, which have to strive hard to deal with the high dimensionality of the design space. This is where design reuse can play an important role.

Table 8.1 Summary of product family design approaches

Approach (* indicates that the approach is aimed at the design of individual products instead of a product family)	Information modeling — Content						Information analysis — Style				Tool / algorithm	Design synthesis — Style			Algorithm	Solution evaluation						Application
	Function	Behavior	Structure (form)	Interface	Taxonomy	Modeling language	None (assumed)	Manual	Heuristic rule	Computational		Configuration design	Manual / case retrieval	Automated synthesis		Objectives	Cost estimation	Qualitative preference	Equation	Meta-modeling	Aggregation function	
Scale-based																						
D'Souza & Simpson [2003]										x	GA			x	GA	M				DOE	x	Aviation aircrafts
Fujita & Yoshioka [2003]										x	LPP			x	LPP	M	x			RSM	x	Lift-gate dumpers
Hernandez et al. [2002]										x	GA			x	GA	M				RSM	x	Universal electric motor
Nayak et al. [2002]										x	SLP			x	SLP	M			x		x	Universal electric motor
Simpson et al. [2001]										x	GRG			x	GRG	M			x		x	Universal electric motor

208 *Design Reuse in Product Development Modeling, Analysis and Optimization*

Category	Reference												Method				Product
Model-based	*Campbell et al. [1999]	x	x	x	x					x	x		SAGA	x			Weighing machine
	Chidambaram & Agogino [1999]	x		x						x	x		GA	x	x		Brushless DC motors
	*Counsell et al. [1999]	x	x	x	x			O-O		x							Factory automation
	*Duffy et al. [1996]	x	x					O-O		x	x						Pump
	*Szykman et al. [1998]	x	x	x	x			XML		ML	x	x		x			Cordless screwdriver
Graph-based	Corbett & Rosen [2004]	x		x		x	GRA			CCP	x	x		PT	x		Automotive underbodies
	Du et al. [2002a]	x		x			GRA			PAGG	x	x		GR	x		Power supplies
	Du et al. [2002b]	x	x	x	x		GRA		x		x	x		GR	x		Office chairs
	Siddique & Rosen [2001]	x	x		x	x	GRA			CCP	x	x		CCP			Coffeemakers
	Dahmus et al. [2000]	x		x					x	MM	x						Cordless screwdrivers
Module-b	Du et al. [2001]	x		x					x		x				x	x	Power supplies

A Product Family Design Reuse Methodology 209

Reference										Method	S/M			Example
Fujita et al. [1998a]	x						x			SQP	S			Aircrafts
Fujita et al. [1999]	x					x	x			SA	S			TV receiver circuits
Gonzalez-Zugasti et al. [2000]	x			x		x	x		x		M	x		Interplanetary spacecraft
Gonzalez-Zugasti & Otto [2000]	x					x	x			GA	S			Interplanetary spacecraft
*Gu & Sosale [1999]		x			SA	x							x	Vacuum cleaner
Martin & Ishii [2002]	x			x										Water cooler
McAdams et al. [1999]	x			x	QFD		x							Tea and coffee makers
Nelson et al. [2001]			FRM		MM	x						x		Nail guns
Ong et al. [2006]	x			x			x			NLP	M	x		Bicycles
Rai & Allada [2003]	x			x			x			ES	SM			Power screwdrivers; electric knives
*Sand et al. [2002]	x	x			MPA	x				AB	M	x		Two-way radio
Stone et al. [2000a]	x			x		x	x							Power screwdrivers

Others										Product
Stone et al. [2000b]	x				x (MM)					Power screwdrivers
Zamirowski & Otto [1999]	x	x	x	x						Xerographic products
Fujita & Yoshida [2001]					x	x (GA,B&B)	M	x		Commercial aircrafts
Jiao & Tseng [2004]			x							Power supply
Jiao & Zhang [2005]					x (ARM)		M	x	x (RSM)	Vibration motors

Note:
AB: agent-based   ARM: association rule mining   B&B: branch-and-bound
CCP: constrained Cartesian product ES: exhaustive search   FRM: Frame   GRA: Graph
GRG: generalized reduced gradient   GR: Graph rewriting   LPP: linear physical programming
M: multiple   ML: machine learning   MM: modularity matrix
MPA: module grouping algorithm   NLP: nonlinear programming   O-O: object-oriented
PAGG: programmed attribute graph grammar   PT: partitioning method   S: single
SLP: sequential linear programming SQP: successive quadratic programming

8.1.2 *Model-based approach*

The model-based approaches emphasize the modeling of existing products. The multiple facets of product information have been carefully dealt with and supported by generic modeling languages. Design catalogs and design repositories have been built to make the information easily accessible by different parties. However, these methods have focused more on the design of individual products than on product families. Therefore, the support to build product architecture is inadequate. Design synthesis has been supported by various techniques. For example, A-Design used an agent-based tool that combined simulated annealing (SA) and genetic algorithm (GA) for intelligent configuration generation [Campbell *et al.*, 1999]. In the NIST design repository project [Szykman *et al.*, 1998], design synthesis was restricted to manual retrieval of product cases.

8.1.3 *Graph-based approach*

The graph-based approaches are distinctive for providing support for product modeling and reasoning. In particular, the graph-based approaches have used graph representations to capture the multiple facets of product information in mathematical forms. Moreover, product architecture building and the subsequent design synthesis have been supported by pre-defined graph grammar. However, the graph transformation operators are restrictive in solution generation. Another restriction is the absence of performance estimation, which can be attributed to the lack of operators to establish the relationships between the design variables and performance.

8.1.4 *Module-based approach*

A number of product family design methods are included in the module-based category. Module-based methods provide extensive solutions to the major issues in product family design. Among them, the function-based product information modeling method has been widely adopted. Moreover, the building of the product architecture has been supported by various analytical techniques, such as heuristic rules or computational algorithms. These can reduce the uncertainties in building the product architecture. Various algorithms have been applied to configuration generation, such as an exhaustive search, linear and nonlinear programming and derivative-free methods *(e.g.*, GA and SA). GA has been advocated by many researchers due to its power in solving large combinatorial problems, typical in product configuration design.

Based on this analysis, product family design must be supported by design reuse in the following aspects.

(1) A complete design process model that incorporates tools to address the major issues in product family design.
(2) Effective information modeling that is characterized by a comprehensive, multiple facet information model and formal representation schemes.
(3) Effective information analysis that supports rapid and intelligent product architecture building.
(4) Efficient solution generation that is supported by automated design synthesis techniques.
(5) Solution evaluation that includes both cost and performance features. Performance evaluation should be supported by a metric to accommodate diverse performance criteria.

8.2 An Integrated Design Reuse Process Model

The PFDR methodology defines systematic procedures to design product family. Design reuse is achieved by identifying reusable product architecture and components, followed by simultaneously designing a set of products using automated solution synthesis and evaluation. The process model includes the steps and tools to implement the design reuse methodology. Basically, it involves three major stages, namely product case modeling, knowledge extraction, and design synthesis and evaluation (Figure 8.1). The functions and associated tools of each stage are discussed next.

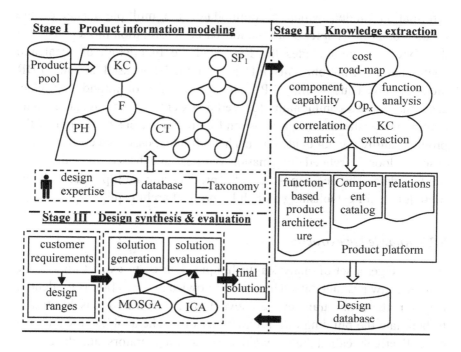

Figure 8.1 The PFDR process model

F: functional PH: physical CT: contextual
SP_i: a general design space Op_x: knowledge extraction operators
ICA: information content assessment KC: key characteristics
MOSGA: multi-objective struggle genetic algorithm

8.2.1 *Product information modeling*

Design reuse begins from the modeling of the product cases that are available from previous designs. The sources of these product cases include past models of the products in a company or those of the competitors. Similar products that share certain commonalities are decomposed within a general design space, denoted as SP_i. Thus, a general design space contains similar products that are expected to form a product family. The decomposition is carried out with respect to a pre-defined information model, which accommodates the multiple facets of the product data. In Chapter 3, a function-based information model has been developed, which includes the KCs, and the functional, physical and contextual information. This information model is used in the PFDR methodology. Thus, within a SP_i, there are a number of individual product cases that share a commonality to a certain extent. At this stage, the innate commonality, the design patterns, and their impact on the design of the product family are not evident yet. The individual cases are considered as a set of loosely related information entities, and hence are called raw product data. The decomposition of a product into multiple facets is the prerequisite for establishing a modular product architecture.

8.2.2 Knowledge extraction

At this stage, a set of analytical techniques is used to extract knowledge from the raw product data. The aim of knowledge extraction is to identify design patterns and transfer the relevant information into reusable forms, to facilitate the building of a product platform. The analytical techniques are collectively called the knowledge extraction operators, and denoted as $\{Op_x\}$. The following operators are developed.

8.2.2.1 Function analysis (Op_f)

Function analysis establishes the function-based product architectures (FPA). The FPA is generalized from the individual function structures that have been built in Stage I. It contains the common functions of a product family, and provides a platform upon which the new products can be built. The SOM method, which has been discussed in Chapter 4, is used to facilitate this process,

8.2.2.2 KC extraction (Op_k)

The KCs that are common to a group of products within the same general design space are extracted. They are used to limit the scope of the problem by focusing on a set of crucial business or customer factors that defines the spectrum of the product family. KC extraction is carried out by senior engineers who are familiar with the market and customer requirements.

8.2.2.3 Correlation matrix (Op_r)

A mapping route from the design parameters (DPs) to the major product performance is established. The DPs are the different attributes of the physical components. As an objective of the design reuse system, the performance of a product family will be optimized with respect to cost and performance. To do so, the performance of the product family has to be estimated based on the attributes of the physical components. Establishing the relationships between the performance of the product family and the DPs is necessary for performance evaluation. In this book, the correlation is established as a few correlation matrices similar to the house of quality (HOQ) in the quality function deployment (QFD) method.

8.2.2.4 Cost modeling (Op_c)

A cost model of the product family is established and the cost road-maps of the components are identified (Chapter 6). Various cost elements can be considered in the cost model. In design reuse, the model includes the cost elements at three levels, namely, the corporation/department level, the product level and the component level. Analytical tools are used to determine these different cost elements. The cost road-maps are established for the components based on existing product cases or quotations from the OEMs.

8.2.2.5 Extracting component capability index (Op_i)

The component capability index refers to the extent to which a component can satisfy the related performance requirements. The capability indices are defined for individual components or component combinations. It is useful in the performance evaluation based on information content. The capability indices are extracted from existing products. Chapter 7 has included the procedures to establish the capability indices and use them to evaluate product performance.

Using these knowledge extraction operators, a product platform is constructed and a component catalog is built. The relationships between the different aspects of the product information are established. These constitute the design knowledge base that can be accessed by different designers.

8.2.3 *Design synthesis and evaluation*

The last stage involves design synthesis and evaluation, in which the candidate configurations of a set of related products are generated, and the production cost and product performance of these configurations are evaluated. Design synthesis and evaluation is formulated as a multiple-objective optimization problem (MOOP), and solved using the multi-objective struggle genetic algorithm (MOSGA) proposed in Chapter 5. Evaluation is performed according to two criteria, namely cost and performance. Cost is estimated using the cost model developed in Stage II. Performance is evaluated as the average information content of a product family. After the optimization process, the final solution can be chosen from the Pareto-optimal set based on the designer's preferences.

The PFDR methodology establishes a consistent framework to facilitate product family design based on the design reuse rationale. It emphasizes the application of AI techniques in the establishment of product platforms and the generation of product family.

8.3 A Web-Based Product Family Design Reuse System

A prototype system was developed to implement the PFDR methodology. Figure 8.2 shows the architecture of this system. This system emphasizes the processing and transformation of product information across different stages of the PFDR process model. The central knowledge base acts as a container of both the raw product data and the refined knowledge obtained through knowledge extraction. It is the source of information for design synthesis. The processing power of the system resides in three core processing engines (the shaded rectangles), namely, the information modeling engine, the knowledge extraction engine and the solution generation engine. These processing engines are supported by various techniques and algorithms. The processing engines accept requests from the designers, such as data entry, knowledge extraction and design synthesis, trigger the corresponding computational tools and communicate with the central knowledge base to retrieve data.

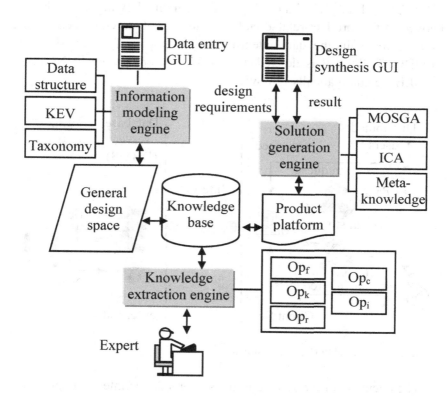

Figure 8.2 Architecture of the PFDR prototype system

The infrastructure of a web-based PFDR prototype system is displayed in Figure 8.3. This system features a client-server structure. The server side hosts the central database and the database management system (DBMS) stores and maintains the product information.

Product data is assorted as a number of interconnected tables. For example, a PRODUCT TABLE contains the records of individual product cases. It is the master table of a FUNCTION TABLE, which is the slave table and contains the records of the functional information. A foreign key (FK) is used to direct a record in the slave table (*e.g.*, the FUNCTION TABLE) to an unique record in the master table (*e.g.*, the PRODUCT TABLE). Similarly, the KC TABLE, the PHYSICAL TABLE, and the CONTEXTUAL TABLE are developed as the slave tables of the PRODUCT TABLE. Figure 8.4 is a diagram showing the relations between a few product-related tables. A number of stored procedures are used to maintain the data integrity. For example, if a record in the PRODUCT TABLE is deleted, the related records in the FUNCTION TABLE are automatically removed.

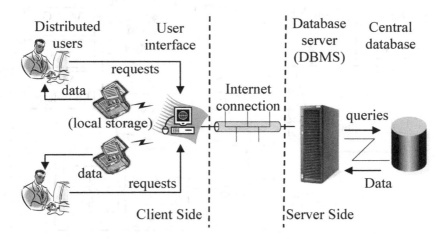

Figure 8.3 Web-based PFDR system structure

The client side is connected to the server side via internet connection. A graphical user interface (GUI) is used to facilitate the communication between an end-user and the server. The data entry GUI is designed to allow for easy entry of product information (Figure 8.5). The multiple facets of the product information can be modeled using pre-defined structures. For example, the function structure of a product can be constructed using the GUI shown in Figure 8.6. With the support of taxonomy, function decomposition can be carried out using simple 'click-and-assemble' operations, which can significantly alleviate the effort to model product functions.

A Product Family Design Reuse Methodology 219

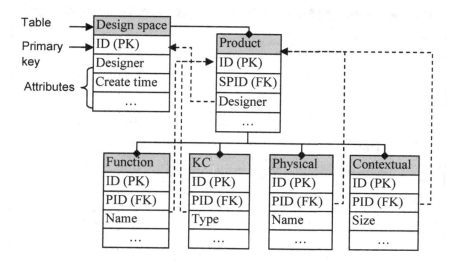

Figure 8.4 Relationship between product-related tables

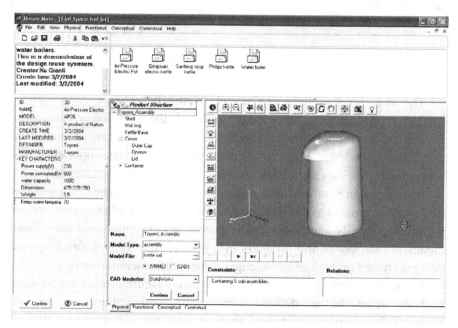

Figure 8.5 User interface for product information modeling

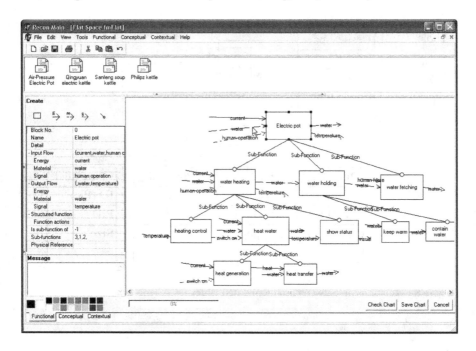

Figure 8.6 User interface for product function decomposition

The raw product data are stored in the knowledge base. When sufficient product cases have been collected, the knowledge extraction engine is triggered to allow an expert to reconfigure the raw product data using the knowledge extraction operators $\{Op_x\}$. These operators are supported by various techniques, either computationally or manually.

At the design synthesis and evaluation stage, a designer can input the design requirements to define a product family. The number of products in the product family and the design requirements are input manually. That means the variety of the products in a product family is defined by the designer based on his/her knowledge of the market segmentations.

Next, the PFDR system will generate the configurations of the product variants using the pre-defined components, and achieve an optimal trade-off between product performance and cost. Automated design synthesis is carried out using the design synthesis engine, which is driven by the MOSGA and the ICA method. MOSGA can cater to any number of products that a designer inputs. However, the algorithm itself does not decide the optimal number of products in the product family. The principles and methods to determine an optimal profile of product variants are beyond the scope of the PFDR methodology. For a more comprehensive study of such issues, readers can refer to Chen *et al.* [1996], Simpson *et al.* [1997a], and Grante and Andersson [2003]. The GUI for design synthesis is shown in Figure 8.7.

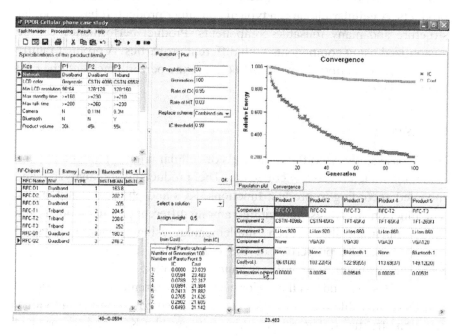

Figure 8.7 User interface for design synthesis

In addition, the system also provides a local storage mechanism by which the frequently visited data are stored temporarily in the local computer. This mechanism allows a user to access the relevant data locally, instead of sending and receiving large volume of data remotely each time. Thus, the network load can be reduced.

Based on the PFDR system, two design tasks have been carried out and are presented as case studies in the next two sections. The design of the cellular phone products shows the effectiveness of the PFDR methodology in automated design synthesis and evaluation. The design of the TV receiver circuits demonstrates the advantages of the design reuse method as compared to the modular design method.

8.4 Design of Cellular Phone Product Family

This case study focuses on the design of a family of cellular phone products. A cellular phone is a typical consumer product. It has a modular structure and many reusable components. Product performance can be customized using components with different properties. A product family is designed as an optimal trade-off between the performance and the cost of the product.

8.4.1 *Settings*

In this case study, 96 different models of cellular phones produced by the same manufacturer were collected. The products were analyzed and modeled according to product functions, KCs and physical components. The function structures were similar among different product models, and many functions have been implemented using standard physical components. The properties of the components, *e.g.*, cost and major technical attributes, were collected and documented.

The KCs that address the major customer requirements were extracted and listed in Table 8.2. An FPA was established based on the individual function structures. The relationships between the KCs and the product functions were established using a correlation matrix, *TR$_2$*, in Table 8.3. This book uses a subset of KCs and functions in the case study for brevity. The component slots that correspond to product functions are shown in Table 8.4. The physical components belonging to the actual product cases are assigned to the component slots.

Table 8.2 KCs of the cellular phones

KCs	Abbr.	Type	Unit	Default value
Network	NW	categorical	–	Dualband, Triband, Quadband
Display color	DC	categorical	–	grayscale, CSTN4096, TFT65k ...
Display resolution	DS	categorical	pixel	96×64, 128×128 128×160, 176×220...
Max talk time	MTT	real	hour	–
Max standby time	MST	real	minute	–
Built-in camera	CM	ordinal	mega pixels	0, 0.11, 0.3, 1.2
Bluetooth connection	BC	Boolean	–	True, False

Table 8.3 Correlation between KCs and functions (TR_2)

KCs		f_1	f_2	f_3	f_4	f_5
k_1	NW	1	0	0	0	0
k_2	DC	0	1	0	0	0
k_3	DS	0	1	0	0	0
k_4	MTT	1	0	1	0	0
k_5	MST	1	1	1	0	0
k_6	CM	0	0	0	1	0
k_7	BC	0	0	0	0	1

f_1: Signal processing f_2: Display f_3: Power supply
f_4: Image capturing f_5: I/O connection

Table 8.4 Function and component slot

Function		Component slot		Number of Components
f_1	Signal processing	s_1	RF chipset & base band	8
f_2	Display	s_2	LCD screen	16
f_3	Power supply	s_3	Battery	7
f_4	Image capturing	s_4	Camera	5
f_5	I/O connection	s_5	Bluetooth set	3

Based on previous discussions, the cost road-maps and capability indices of the components have to be established. Basically the cost road-maps are available from market analysis and historical data. Since such information is proprietary and highly confidential, the actual cost data was not used in this book. Instead, the cost road-map was simulated using a simplified model, given in Equation (8.1).

$$C_c^j = C_c^0 \left(\frac{V_j}{V_0}\right)^{R_j} \tag{8.1}$$

where C_c^j is the cost of component j, C_c^0 is the cost at a given production volume V_0, V_j is the volume of component j used in the product family. R_j is a regression coefficient that simulates the learning effect of production. $R_j = \log_2 s$, where s is the slope of the learning curve. Typically in the electronics industry, $90\% \le s \le 95\%$, and hence, $-0.15 \le R_j \le -0.074$.

The cost at the department level was fixed. Moreover, the cost at the product level was computed according to Equation (6.5) based on the complexities of the different products.

The component capability indices that correspond to the KCs, namely, *NW*, *DC*, *DS*, *CM* and *BC* were easily established because each KC is related to only one function. On the other hand, the *MST* and *MTT* are related to more than one function. In particular, *MST* is related to three functions, namely, f_1 (signal processing), f_2 (display) and f_3 (power supply), and *MTT* is related to two, namely, f_1 and f_3. The capability indices of these two component combinations were established using statistical methods. For example, with respect to k_4=*MTT*, the component combination (RFC-T1, Li-Ion 860) has been used in five host products $\boldsymbol{h} = \left[h^1, h^2, h^3, h^4, h^5\right]$. Ten samples were retrieved for each host product and the *MTT* values of these samples were available from historical data (Table 8.5). The capability index was defined as a normal distribution, $\{(\text{RFC-T1,Li-Ion 860})|MTT: \mu = 226, \sigma = 25.6\}$. Similar procedures were carried out for the other component combination.

Six products are to be launched as a product family. The design requirements are listed in Table 8.6, with the last row indicating the production volumes. The MOSGA method was used to generate the optimal product family. After a set of Pareto-optimal solutions were generated, different weights were assigned to the two objectives, namely, cost and information content.

Table 8.5 Sampled MTT values of five products that host (RFC-T1, Li-Ion 860)

Sample	h^1	h^2	h^3	h^4	h^5
1	249	209	273	242	216
2	238	237	233	194	225
3	260	189	254	255	237
4	237	228	236	208	201
5	222	205	275	196	201
6	218	246	259	227	240
7	230	210	227	184	216
8	250	224	253	234	219
9	251	231	239	209	242
10	228	224	247	221	223

Table 8.6 Design requirements of the product family

KCs		P_1	P_2	P_3	P_4	P_5	P_6
k_1	NW	Dualband	Dualband	Triband	Triband	Triband	Quadband
k_2	DC	grayscale	CSTN4096	CSTN65535	TFT65535	TFT260K	TFT65535
k_3	DS	96×64	128×128	120×160	128×160	176×220	176×220
k_4	MTT	≥160	≥230	≥210	≥200	≥170	≥180
k_5	MST	≥200	≥260	≥230	≥250	≥200	≥190
k_6	CM	0	0.11M	0.3M	0.3M	1.2M	1.2M
k_7	BC	False	False	True	False	True	True
Production volume		30,000	45,000	55,000	37,000	20,000	15,000

Information content is used to measure the performance of a product, as discussed in Section 6.3. Thus, the product family could accommodate the designer's preference on reducing cost or enhancing performance. For example, if a larger weight is assigned to the information content, the selection of the optimal solution is more sensitive to variations in the information content. As a result, a solution with a smaller information content (which means a better performance) is selected.

In this case study, w_1 is the weight assigned to performance (information content), and $w_2=1-w_1$ is the weight assigned to cost. Three strategies were adopted, namely performance priority ($w_1=0.8$, $w_2=0.2$), cost priority ($w_1=0.2$, $w_2=0.8$), and equal priority ($w_1=w_2=0.5$).

8.4.2 *Results*

Figure 8.8 illustrates the objective functions of three solutions according to three priority strategies. Tables 8.7, 8.8 and 8.9 show the product configurations generated using the PFDR method according to these three priority strategies. The last row in each table shows the objective functions, and the last column shows the number of different components used in a product family.

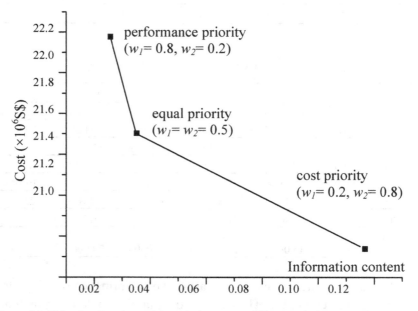

Figure 8.8 Objective functions of the solutions w.r.t. different priority strategies

Table 8.7 Product configurations (performance priority)

	P_1	P_2	P_3	P_4	P_5	P_6	#
s_1	RFC-D3	RFC-D2	RFC-T2	RFC-T2	RFC-T3	RFC-Q2	5
s_2	Grayscale	CSTN-4096b	CSTN-65Kc	TFT-65Kd	TFT-260Kf	TFT-65Ke	6
s_3	Li-Ion 780	Li-Ion 860	Li-Ion 700	Li-Ion 860	Li-Ion 600	Li-Ion 780	4
s_4	None	VGA11	VGA30	VGA30	VGA120	VGA120	3
s_5	None	None	Bluetooth 1	None	Bluetooth 1	Bluetooth 1	1
IC	0.000	0.033	0.001	0.000	0.116	0.008	–
Cost	71.94	89.90	116.32	116.03	150.06	152.80	–
Average IC: 0.026	Total cost: 22.188($\times 10^6$S$)						19

Table 8.8 Product configurations (equal priority)

	P_1	P_2	P_3	P_4	P_5	P_6	#
s_1	RFC-D3	RFC-D2	RFC-T2	RFC-T2	RFC-T2	RFC-Q2	4
s_2	Grayscale	CSTN-4096b	CSTN-65Kc	TFT-65Ke	TFT-260Kf	TFT-65Ke	5
s_3	Li-Ion 860	Li-Ion 780	Li-Ion 780	Li-Ion 780	Li-Ion 780	Li-Ion 780	2
s_4	None	VGA30	VGA30	VGA30	VGA120	VGA120	2
s_5	None	None	Bluetooth 1	None	Bluetooth 1	Bluetooth 1	1
IC	0.000	0.0326	0.003	0.166	0.000	0.007	–
Cost	75.15	92.77	113.60	109.50	139.77	145.03	–
Average IC: 0.035	Total cost: 21.7($\times 10^6$S$)						14

Table 8.9 Product configurations (cost priority)

	P_1	P_2	P_3	P_4	P_5	P_6	#
s_1	RFC-D3	RFC-D2	RFC-T2	RFC-T2	RFC-T2	RFC-Q2	4
s_2	CSTN-4096b	CSTN-4096b	CSTN-65Kc	TFT-65Ke	TFT-260Ke	TFT-65Ke	3
s_3	Li-Ion 600	Li-Ion 600	Li-Ion 780	Li-Ion 780	Li-Ion 780	Li-Ion 780	2
s_4	None	VGA30	VGA30	VGA30	VGA120	VGA120	2
s_5	None	None	Bluetooth 1	None	Bluetooth 1	Bluetooth 1	1
IC	0.000	0.525	0.003	0.166	0.000	0.007	–
Cost	75.15	83.25	112.59	108.50	138.76	144.03	–
Average IC: 0.117	Total cost: 21.144($\times 10^6$S$)						12

8.4.3 *Analysis*

The cost of the product family based on cost priority was 21.144×10^6S\$, which was about 5% lower than the cost based on performance priority (22.188×10^6S\$). The cost differences are attributable to the different levels of commonality, which can be roughly estimated as the number of different components implemented in the product family. With respect to the different priority strategies, the numbers of different components were 19 (performance priority), 14 (equal priority), and 12 (cost priority), respectively. This indicates that cost effectiveness has been achieved by increasing the commonality among the products in the family.

The average information content was used to measure the performance of the product family. The minimum average information content (0.026) was found when performance was given a higher priority ($w_1=0.8$). The product family in this scenario could fulfill the design requirements to a higher extent. Based on Equation (7.4), the probability that the design requirements can be satisfied is 98%.

It should be noted that the average information content is used for the purpose of optimizing the entire product family. It may not reflect the performance of individual products. Instead, the information content of an individual product indicates the performance of that particular product. For example, the information content of product P_2 in Table 8.9 is 0.525, which means that the probability that the design requirement can be satisfied was only 70%. The main reason for this inferior performance is that the KCs, namely, *MST* and *MTT*, were poorly satisfied. Studying the components of this product, a low capacity battery has been the cause of the poor *MST* and *MTT* performance. This provides guidelines to the designer to improve the solution, *e.g.*, replacing the low capacity batteries with high capacity ones.

This case study shows the design synthesis and evaluation process of product family design. Cost and performance considerations can be effectively managed for product family design within the optimization framework. Information content is used as an integrated measurement of product performance. The estimation of product performance based on information content can also be used to identify possible flaws of a design, and accordingly, provide guidelines for improvements.

8.5 Design of TV Receiver Circuits

This section presents a comparative study of the design of a family of TV receiver circuits. In particular, Fujita *et al.* [1999] proposed a modular method for product family configuration design, which is used as a comparison benchmark. Using the same set of data, the PFDR method is used to solve the same problem. Results obtained from this method were compared to those from the benchmark method. The effectiveness of the PFDR method is discussed.

8.5.1 *Settings*

The target problem is the same as in the benchmark method, *i.e.*, to design a family of six TV receiver circuits. Two scenarios were considered, namely, case 1 and case 2, in which the production volumes are different (Table 8.10).

The key characteristics of the product (called **feature indices** in the benchmark method) were kept the same (first column of Table 8.11).

The product architecture, as well as the functions and product modules, remained unchanged (Figure 8.9). In particular, seven module slots were reused. Based on the original relationships between the KCs, functions and the modules, a correlation matrix was easily obtained (Table 8.11).

Table 8.10 Product variety of TV sets

KCs		P_1	P_2	P_3	P_4	P_5	P_6
Picture size (inch)		14	21	36	14	21	36
Picture quality		Good	Better	Best	Normal	Good	Better
Audio level		Low	Medium	High	Low	Medium	Medium
Power supply voltage		100V	100V	100V	Multi	Multi	100V
Production volume	Case 1	36000	24000	12000	36000	24000	12000
	Case 2	24000	24000	24000	24000	24000	24000

230 Design Reuse in Product Development Modeling, Analysis and Optimization

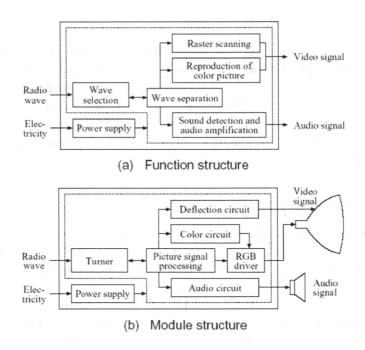

Figure 8.9 Function and modular diagram of a receiver circuit [Fujita et al., 1999]

Table 8.11 Mapping from design requirements to components

KCs	M^1	M^2	M^3	M^4	M^5	M^6	M^7
(Defaults)	1						
Picture size			1				
Picture quality		1		1	1		
Audio level						1	
Power supply voltage							1

M^1- Tuner $\qquad$ M^2- Picture signal processing
M^3- Deflection circuit $\qquad$ M^4- Color circuit
M^5- RGB driver $\qquad$ M^6- Audio circuit
M^7- Power supply

A Product Family Design Reuse Methodology 231

To satisfy the product performance requirements, the PFDR method used the information content, instead of a set of design constraints, which was the case in the benchmark method. In order to do so, the capability indices of the different components must be established. The capability indices were extracted from the attributes of the components. As an example, the capability indices of three deflection circuits (M^3) are presented next.

Module M^3 corresponds to the KC, *picture size*, which is an ordinal type KC and the default values are 14-inch, 21-inch and 36-inch. Three components m_1^3, m_2^3, and m_3^3 were available. m_1^3 can be used for the 14-inch picture size only, while m_2^3 can be used for both the 14- and 21-inch picture sizes. m_3^3 can be used for all the three picture sizes. The capability indices of these three components were established as the pmfs shown in Figure 8.10. Similarly, the capability indices were established for the other modules. It should be noted that, in the benchmark method, the 'capacity constraint' was used to deal with the interrelationships between the modules. This constraint was retained in the PFDR method to define the feasible design space.

For the product family cost, the PFDR method uses the cost model proposed in Chapter 6. To use this model, the cost elements and detailed data in the benchmark method was reformulated. Firstly, the original cost elements were assigned to three levels; next, the cost road-maps were established for each component based on the original data. The cost model in relation to the original cost elements in the benchmark method is shown in Table 8.12.

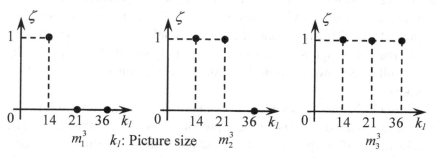

Figure 8.10 Component capability indices of three deflection circuit components

Table 8.12 Cost model reformulation based on Fujita *et al.* [1999]

	PFDR cost model	Original cost element
Department level	$C_F = C_f^0 = 10{,}000{,}000$	$C_f^0 = 10{,}000{,}000$
Product level	$\sum C_P^i = C_v^a + C_f^P = 9{,}630{,}703$	$C_v^a = 645703,$ $C_f^P = 8985000$
Component level	$C_{c(k)}^j = c_0 N_k^j \left(\dfrac{n_k^j}{n_0} \right)^{R_j}$ where: $c_0 = 1.8$, is the average cost per primitive. N_k^j is the number of primitives in component m_k^j. n_k^j is the number of component m_k^j in the product family. $n_0 = 24000$ is a base number of component. R_j is the learning regression exponent, which is defined the same as in Equation (8.1).	Fixed cost: $\sum C_f^{P(i)}$; $\sum C_f^M$; Variable cost: C_v^m - material cost; C_v^f - facility cost.

8.5.2 *Solution generation and results*

In the PFDR method, design synthesis was formulated as a MOOP and solved using the MOSGA method. Product performance was evaluated according to the information content. The algorithm generates the solution that minimizes the product family cost and the average information content. The information content of the optimal solutions is zero, which means that the design requirements can be safely satisfied. Product family cost results are shown in Table 8.13. Results of the benchmark method are reproduced.

It should be noted that in the benchmark method, a primal solution is generated without considering product family optimality. The configuration of this primal solution and the resulting product family cost are compared with those of the optimal solutions, in case 1 and case 2 respectively. This strategy is also adopted in this book.

Table 8.13 Optimization results of product family cost

		Fujita's method			PFDR method		
		Primal	Optimal	diff. (%)	Primal	Optimal	diff. (%)
Case 1	# of diff. modules	20	11	−45%	20	11	−45%
	Total cost (10^6 Yen)	133.112	130.279	−2.1%	133.317	131.219	−1.6%
Case 2	# of diff. modules	20	11	−45%	20	10	−50%
	Total cost (10^6 Yen)	135.251	131.759	−2.6%	135.919	132.396	−2.4%

8.5.3 *Comparison*

The configurations of the optimal solutions based on the PFDR method and the benchmark method were similar to each other. As compared with the primal solution, both optimal solutions reduce the number of components by 45 ~ 50%, achieving the same level of commonality. Since a higher commonality is key to cost reduction, the optimization method has effectively incorporated component commonality to reduce cost.

The total product family costs are of almost the same value for both the PFDR method and the benchmark method. This is an indication that the cost model of the PFDR method has reflected the essential cost elements. In the PFDR method, the cost reduction is 1.6% in case 1, and 2.4% in case 2. In the benchmark method, the optimal solutions achieve cost reductions of 2.1% and 2.6%, respectively. From the results, the benchmark method achieves a higher cost reduction. However, this is not an indication that the PFDR method is inferior in reducing cost. The differences can be attributed to the different formulations of the cost model. Although the cost model of the benchmark method is very comprehensive, it involves a number of parameters and coefficients that are difficult to be assigned accurately and reliably [Fujita, 2002]. For example, the coefficients were assumed based on references rather than the actual production practices of the specific products. This is attributable to the ambiguity to forecast the design and manufacturing processes at the design and planning phase.

The performance requirements have been satisfied in both methods. In the benchmark method, this was achieved using three sets of design constraints which are problem specific. If the design requirements are changed, or the properties of the modules are modified, the constraints and optimization procedures have to be changed accordingly. Moreover, the problem formulation is applicable to discrete KCs only, and presents difficulties in dealing with a mixture of discrete and continuous KCs. In the PFDR method, performance is ensured through minimizing the information content. The design requirements and the module properties can be changed without affecting the design synthesis and the evaluation process. Discrete and continuous KCs can be dealt with consistently in the ICA method. These have been possible because the ICA method can consistently define the capability indices and compute the information content based on these indices. Thus, the PFDR method provides a more generic formulation to ensure product performance.

The major advantages of the PFDR method in comparison with the benchmark method are summarized in Table 8.14.

Table 8.14 Comparison of the PFDR method and the benchmark method

	Benchmark method	PFDR method	Comment
Problem formulation	Constraint-based single-objective optimization.	Multi-objective optimization in a comprehensive design reuse framework.	PFDR provides a more generic formulation based on the design reuse methodology.
Product performance	Based on three sets of design constraints, which are problem specific.	The ICA method; Component capability indices and mapping from KCs to components derivable from existing cases.	The ICA method is applicable to different types of KCs. It presents standard procedures to compute the information content for performance evaluation.
Cost model	Based on detailed design and production information; Various parameters are assigned arbitrarily.	Based on a three-level cost model derivable from historical data.	Component cost road-map is used in the PFDR method to estimate cost. Relevant information can be obtained from existing designs.

8.6 Summary

This chapter is an extension and summary of previous discussions. Based on the belief that a unified process model is more powerful in facilitating decision-making for product family design, a PFDR methodology is developed to integrate various design reuse techniques into a unified model. This methodology represents an effort to combine the individual design reuse techniques, so that they can contribute to product design from various aspects. A web-based prototype system is developed to implement this methodology and two case studies are presented to show the power of this system. It was demonstrated that the system has the following merits:

(1) The entire design reuse processes can be dealt with using the PFDR system. In particular, the system provides a comprehensive solution to domain exploration, knowledge extraction and design-by-reuse.

(2) Automated design synthesis can be achieved with an efficient exploration of the large design space of configuration design.

(3) The evaluation enjoys the use of information content as a uniform metric for product performance. Hence, solution evaluation is carried out along with the design synthesis process so that the design process is very efficient.

Chapter 9

Design Reuse for Embodiment and Detailed Design

Previous chapters have focused mainly on the conceptual stage of product design, which determines the schematic solution principles of products. Although conceptual design is the most important factor for determining product value and investment return, the subsequent embodiment and detailed design stage should not be overlooked because without these efforts, the product could not be realized. This chapter presents an online Web-based environment for product embodiment and detailed design. In particular, case-based reasoning (CBR) is used to implement the embodiment design based on two search methods, namely, the case similarity metrics and a GA-based optimal search. An adaptive design transformation method is applied for detailed design. A Web-based design reuse system is developed to implement the proposed methodology.

9.1 Introduction

Embodiment design is the part of a design process in which a design is developed considering technical implementations and economic criteria. Subsequently, detailed design will finalize the details of the product model, such as geometry, tolerance, material, etc., which lead directly to production [Pahl and Beitz, 1996]. Embodiment design follows a set of rules to ensure the fulfillment of the technical functions, technical and economic feasibility and ease of production. The techniques to facilitate the process include Design for Manufacture and Assembly (DFMA) [Alting and Legarth, 1995], design for disassembly [Jovan et al., 1993; Johnson and Wang, 1998], design for environment [Allenby, 1991], Failure Modes and Effects Analysis (FMEA) [Hata et al., 1997], etc.

The detailed design phase is so far the most explicitly understood and computerized phase in the design process as compared with other phases such as conceptual design or embodiment design. Related research issues include detailed design data storing methods [Burgess et al., 1998; Andrews and Sivaloganathan, 1998], database construction [Regliand and Cicirello, 2000], feature extraction techniques and geometrical data similarity evaluation [Regliand and Cicirello, 2000], development of standard languages for detailed design description and product modeling [Brunetti and Golob, 2000], functionality capturing of detailed design data [Burgess et al., 1998; Jensen, 1998], and retrieval of the detailed design data [Altmeyer and Schurmann, 1996]. These efforts, to some extent, embody Duffy's design reuse model, namely design-by-reuse, design-for-reuse and domain-exploration [Duffy et al., 1996]. For example, functionality capturing of the detailed design data, which is conducted during product design, is a form of design-for-reuse; and the functionality recorded can be used as a search index for product data retrieval (design-by-reuse); design description languages and the methods for storing and for database construction are designed to facilitate product representation, recognition, information extraction and retrieval, which are issues falling into the scope of domain exploration.

Despite progress in this area, much of the research and many of the prototype systems are implemented on a single computer or workstation as a standalone system based on a certain solid modeler [Duffy et al., 1998]. However, the design realization process has become more globalized and distributed, and other product development life-cycle activities, such as marketing, manufacturing, etc., have become tightly interwoven with design [Zhang and Xue, 2001]. Under these circumstances, a standalone system obviously cannot fulfill this need, and an open environment that can be easily accessed to assist product design by reusing previous design is of great importance and benefit. It will greatly promote the understanding and utilization of previous designs, as well as the evolution of new design.

9.2 Online Design Reuse System

This chapter is concerned with developing a Web-based working system to facilitate embodiment design and detailed design. Detailed implementation issues are presented to demonstrate how the design processes can be computerized and facilitated by the Internet technologies. This section discusses the overall structure of the system. The implementation details of the embodiment design and detailed design are discussed in Sections 9.3 and 9.4, respectively.

9.2.1 *System architecture*

The design reuse system presented in this chapter consists of two layers, namely (1) the embodiment design layer based on various case search methods, and (2) the detailed design layer based on detailed design transformation techniques (Figure 9.1).

Basically, the embodiment design layer involves the search for appropriate solutions from the design case base. Two modules are defined, namely the simple search module and the advanced search module. The simple search module retrieves reusable products using an enhanced database enquiry method. As existing data have been properly indexed, they can be retrieved based on these indices. The implementation of the simple search method is straightforward and hence is not elaborated in the book. The advanced search module provides more intelligent retrieving mechanisms. In this module, there are four sub-modules, which are discussed next. The input to this module is the desired function requirement for a new product.

Function analysis module. Since descriptions in natural language are highly diversified and not sufficiently robust to be used as a retrieving index, they will be translated into a formal format by a function analysis sub-module so that they can be encoded and processed computationally.

Product family retrieval module. With the analyzed function requirement, the product family retrieving sub-module will search for the existing product database for a product family with the most similar function. Since a product family usually consists of many member products, the product retrieving sub-module further explores this product family to find a most similar product to reuse.

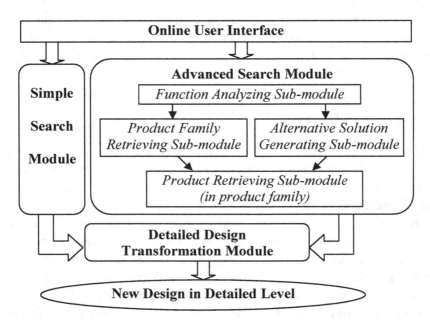

Figure 9.1 Framework the of the design reuse system

Alternative solution generating module. In cases where no similar product family is available, the system combines a few existing products to develop a new design. The alternative solution generating sub-module simulates this process to some extent. It will search in the database for alternative solutions to meet the input function requirement.

Product retrieving sub-module. Several possible solutions will be presented and ranked according to criteria such as cost or weight. The product retrieving sub-module is executed to identify the most possible product to reuse.

The output of the embodiment layer is a reusable product. Next, the detailed design transformation module will generate the new design at the detailed design level based on the retrieved product. At this point, a complete design reuse process from an abstract function requirement to a concrete new detailed design is completed.

9.2.2 Product information representation

To accommodate the requirements of embodiment and detailed design, the product information model is defined based on three major elements, namely the product structure hierarchy, the product profile and the product function. In the product hierarchy, the product structure is presented in a tree-like format, where a branch represents an assembly, and a leaf node represents a single part. The product profile contains the important information of the elements in the product hierarchy, including the key characteristics (KCs) and geometric models. The product function is a function structure showing the flow of energy, material and signal. The function structure is similar to the structure presented in Chapter 3. However, for simplicity of discussion, the hierarchy has two levels, namely the branch-level denoting the main functions and the leaf-level denoting the sub-functions. In addition, the function taxonomy is slightly different from the previous structure presented in Chapter 3. In particular, the taxonomy is based on the work reported by Kirschman and Fadel [1998].

9.3 Embodiment Design

Embodiment design is responsible for retrieving a product or product family with respect to the desired functions. The system provides two methods to solve a design problem. The first method searches for a single product family for reuse while the second method searches for a combination of several existing product families as a solution to the problem.

9.3.1 *Product case retrieval method*

In design reuse, there is a reasoning process that guides users from the original design problem to a reusable design. In this book, the framework partially employs the CBR concept and conducts design reuse using CBR procedures, such as problem clarification, case retrieval and case adaptation. In CBR, cases are compared to evaluate their similarity based on similarity metrics. The design of the similarity metrics varies for different types of cases. However, the underlying methodologies can be summarized into two categories, namely the ratio of AND to OR attributes, and the distance, *e.g.*, Manhattan, Euclidean, etc. [Sarker and Islam, 1999]. Both categories of similarity metrics are adopted in this book according to the nature of data handled at certain stages.

9.3.1.1 Similarity metric for product family retrieval

The input to the product family retrieving sub-module is the function description in the form of a function taxonomy, and this sub-module searches the product database for a product family that has the most similar function. Hence, a similarity metric is developed to determine the similarity of the input function with those of the existing products.

The function requirement in the form of a function taxonomy is represented in a tree-like format, with the main functions as the branches and the sub-functions as the leaves. Hence, a general metric given in Equation (9.1) is formulated to determine the similarity of two trees, that of the input function requirement and the function of an existing product being compared with the desired input function. A tree representing a function is assumed to have one level of branches and one level of leaves.

$$SIM\_PF(Tree1, Tree2) = \frac{V(AND(Tree1, Tree2))}{V(OR(Tree1, Tree2))} \quad (9.1)$$

The similarity metric *SIM_PF* between *Tree*1 and *Tree*2 is calculated using Equation (9.1). The basic principle is to determine the ratio of *AND*(*Tree*1, *Tree*2) to *OR*(*Tree*1, *Tree*2) . *AND*(*Tree*1, *Tree*2) is a tree made up of the branches and leaves that are common to both trees. $OR(Tree1, Tree2)$ is a tree made up of all the branches and leaves in both trees. The calculation procedures of this similarity metric are illustrated as follows.

Input:

Step 1 $OR(Tree1, Tree2)$ generates a tree made up of all the branches and leaves in both trees.

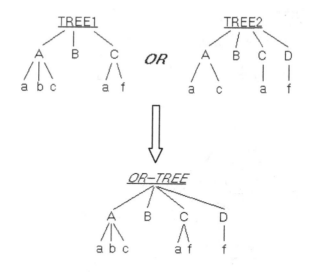

Step 2 $V(OR(Tree1, Tree2))$ quantifies the tree resulting from Step 1, and calculates the value of this tree. Each branch has one point. If there are

n leaves for a branch, each leaf under that branch is valued as $1/n$. The total value of the tree is the sum of the values of all the branches and leaves.

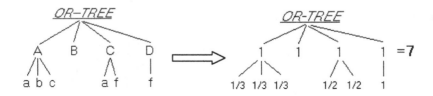

Step 3 *AND(Tree1,Tree2)* generates a tree made up of the branches and leaves that are common to both trees.

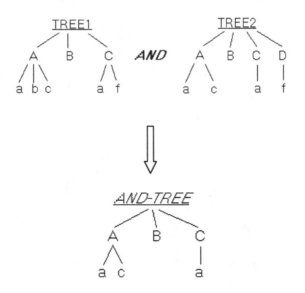

Step 4 $V(AND(Tree1, Tree2))$ quantifies the tree resulting from Step 3 in the same way as in Step 2. However, each branch and leaf carries the same value with the corresponding one in $OR(Tree1, Tree2)$. The value of the tree is the sum of the values of all the branches and leaves.

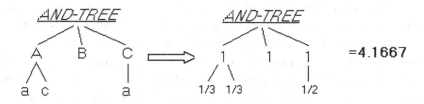

Step 5: The similarity *SIM_PF* between the two trees is calculated.

$$SIM\_PF(Tree1, Tree2) = \frac{V(AND(Tree1, Tree2))}{V(OR(Tree1, Tree2))} = \frac{4.1667}{7} = 59.52\%$$

With this similarity metric, the similarity between two function trees can be determined. In the product family retrieving sub-module, this metric is used to compare the input function tree with the function trees of all the product families in the database. These trees are ranked based on their similarity metrics with the first having the highest similarity.

9.3.1.2 Similarity metric for product retrieval

When a product family has been retrieved, the information of the product family and the details of the family members are presented in the search results. With this information, users still need to select a product among these family members to reuse. At this step, the users will be prompted to input the desired values for the KCs and other major design targets, such as cost or weight. The module will use these values to determine the similarity metric between these inputs and each member product, and rank them for the users' reference.

KCs can be either qualitative or quantitative. For example, for qualitative KCs such as 'input current type' and 'speed limiter', 'D' indicates direct current, and 'Y/N' indicates whether a product has a speed limiter. Before evaluating the similarity metric, a filter is used to screen out all those member products that do not satisfy the qualitative attributes. Only the quantitative attributes are used to evaluate the similarity metric.

If the user inputs n quantitative values for the intended design, these values can be assumed as a point in the n dimensional coordinate space. The family members in a product family can also be regarded as different points in this space. Equation (9.2) is used to calculate the similarity metric based on the Euclidean distance of two points. First, the distance between the point of the intended design and that of a family member is calculated. Since the units might be different for different attributes, they cannot be added up directly. Hence, the average value of all the members for each attribute, calculated using Equation (9.3), is used as the distance in each coordinate axis of the space. Thereafter, the distances in all the coordinate axes are added up to obtain the distance between the points of the intended product and the member product. Finally, the distance is inverted to obtain the similarity metric. If the distance is zero, Equation (9.2) is not applied. Since a zero distance means the product completely matches the user input, it is ranked first in the results generated by the system.

$$SIM\_P(P, M_i) = \frac{1}{\sqrt{\sum_{j=1}^{n} \frac{|V_{jp}^2 - V_{ji}^2|}{Avg_j}}} \quad (9.2)$$

$$Avg_j = \frac{\sum_{k=1}^{m} V_{jk}}{m} \quad (9.3)$$

where $SIM\_P(P, M_i)$ is the similarity between products P and M_i, where P is the desired product, and M_i is the i th member product in the product family; n is the total number of quantitative attributes calculated, which is also the number of quantitative values of the user inputs; V_{jp} is the value of the j th quantitative attribute of product P; V_{ji} is the value of the j th quantitative attribute of product M_i; Avg_j is the average of the j th quantitative attribute of all the member products in the product family; V_{jk} is the value of the j th quantitative attribute of the k th member product in the product family; and m is the total number of member products in the product family.

9.3.2 *Optimal search for alternative solution*

The alternative solution generating sub-module provides another choice for users to generate new design based on existing products. In particular, this sub-module generates possible solutions by combining several existing products to address the input function requirement. In essence, each existing product may satisfy one or more function entries in the in/out function requirement. Together, they may satisfy the entire requirement.

The solution method of this module is basically an optimization search process. Two optimization methods are available and the method to be used is decided by the computational load for the design problem. When the computational load is not too high, an exhaustive search is used; otherwise, the GA-based search method is employed. Regardless of the search method used, the problem representation, solution formulation, rules to evaluate the validity of a solution and the optimization objective are the same.

9.3.2.1 Encoding of input function requirement

Figure 9.2 illustrates how an input function requirement, after being analyzed, is encoded. An analyzed function requirement is a tree-like data structure with its main functions as branches and sub-functions as leaves under the main functions. To encode an input function requirement, each natural language sentence is translated into a string, such as M1100, according to Table 9.1. The first letter represents the function categories; the following digits are the details. The letters or numbers in the following parentheses are the encoded formats of the words. For the first and second function categories, the fourth digit is set as '0' by default. The signs '@' and '*' are used to distinguish between a main function and a sub-function. The encoding process simplifies the data to be stored and facilitates the implementation of the optimization search methods.

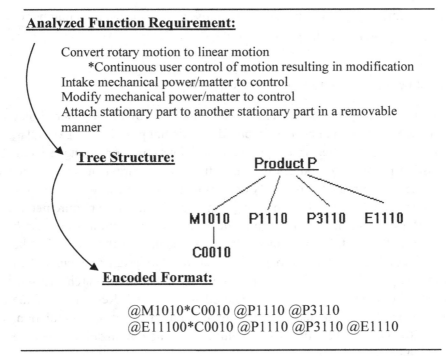

Figure 9.2 Encoding of the function requirement

9.3.2.2 Generating possible solutions

With an input function requirement encoded, the possible solutions can be generated. The requirement in the string in Figure 9.2 '@M1010*C0010@P1110@P3110@E1110' has five sections. To generate a solution based on it, five product families will be retrieved with each product family satisfying one of the function sections. For the sections in the string that represent the main functions, the retrieved product family should also have such a section as its main function. For a section that represents a sub-function, *e.g.*, '*C0010', its corresponding product family should either have such a function as its main function, or it should have such a sub-function under the same main function just as in the encoded input function, *e.g.*, @M1010*C0010. For the five product families retrieved, they may be totally different, or identical for one or more sections.

Figure 9.3 shows several possible results for the string in Figure 9.2. Each result is a combination of several product families. The functions of the component product families are also listed. For the section '*C0010', a product family PF23 with C0010 as its main function is retrieved in the first solution. In the third solution, the retrieved PF11 has '*C0010' as its sub-function under the main function of M1010, which is similar to the input string. In the first solution, all the five product families retrieved are different. In the second and third solutions, four product families are retrieved since PF251 and PF11 can satisfy two sections of the function requirement.

9.3.2.3 Evaluating solution validity

When possible solutions have been generated with several product families combined together, it is possible that these components are not compatible with each other. Thus, the solutions might not be valid. For a product to be valid, its components should be compatible in every aspect, such as geometrical structures, parameters, energy and motion flows. For example, a shaft with $\Phi 10$ diameter can be coupled with another part with a $\Phi 10$ hole; a part outputting rotary motion can be assembled with another component that is driven by rotary motion.

250 *Design Reuse in Product Development Modeling, Analysis and Optimization*

However, at this step, only product families with qualitative attributes are available, and it is impossible to check the compatibility of the components combinations at the feature or parameter level. In the profile of a product family, there are three categories of I/O, namely energy, motion and material. Each category of I/O consists of several types. Table 9.2 lists the details in each I/O category. All the products in one family have the same I/O attributes. Since the I/O information of the product families has been stored in the database, it can be used to estimate the compatibility of the components qualitatively, and hence, the validity of the generated solutions. In the developed system, a simple mechanism is employed to evaluate the validity of a solution at the qualitative level.

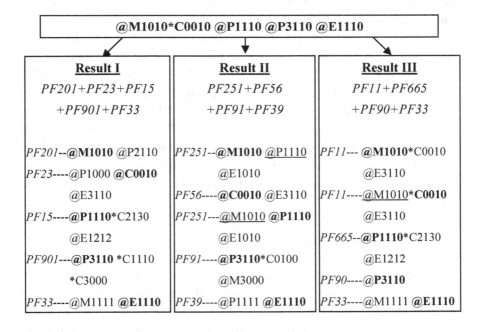

Figure 9.3 Possible solutions

Table 9.1 Encoding of a function requirement [Kirschman and Fadel, 1998].

[Motion]			
Create	Rotary		
Convert	Linear	Motion to	Rotary
Modify	Oscillatory		Linear
Transmit	Other		Oscillatory
Dissipate			Other
[Power/Matter]			
Store	Electrical		Control Heat
Intake	Mechanical	Power/Matter	Move
Expel	Other	to	
Modify			
Dissipate			
[Control]			Power
Continuous	User	Control of	Motion	Resulting in	Modification
Discrete	Feedback		Information		Indication

[Enclosure]			
Support			
Attach			
Contact	a part to a	Moving / Stationary	
Guide			
Limit		Moving / Stationary part in a	Removable / Permanent Matter
Cover			
Protect			
View			

Design Reuse for Embodiment and Detailed Design

Table 9.2 Product/Family I/O Classifications

Energy:	mechanical, electrical, thermal, magnetic, others
Motion:	linear motion, rotary motion, oscillatory motion, others
Material:	solid, fluid, gas, others

In a real product, the energy, material and signal can form a flow. In this system, the retrieved solutions are checked to determine whether the component product families can form a serial energy, motion or material flow using their I/O information. For example, for a solution with n components, it can only be valid in energy flow when these n components can form a serial energy flow. To form a serial energy flow, the output of one component should be of the same type as the input of the following component. The mechanism for checking the motion flow and material flow is the same.

It should be noted that real design problems and products might be very diversified and complicated, and situations such as multi-I/O, flow divergence and convergence are more common than the serial flow situation. Hence, the validity checking procedure is optional in this system. The user can decide whether or not to use this checking mechanism according to the application. In addition, the three types of checking (energy, materials and motion) procedures can be used separately or together to evaluate the validity of a solution depending on the nature of the design problem. This evaluating method is still an experimental effort. To achieve a more accurate evaluation, more in-depth research in this topic is required.

9.3.2.4 Optimization goal

Almost all the product life cycle activities focus on cost reduction. Hence, minimum cost is usually an important target in design activities. However, in certain design domains, such as the automotive or the aeronautic industries, the weight of a product may be more important than the cost of the product. Hence, the alternative solution generating sub-module supports the search for an optimal solution by minimizing either the total cost or weight of the new design. Through the system user interface, the user can select the optimization goal to use.

The cost or weight of a solution is determined by summing all the cost or weight of each of the component product families. Identical product families will only be counted once. Since a product family contains several member products, the cost or weight of the product family refers to the average cost or weight of all the member products. This average cost largely reflects the manufacturing methods and processes of the product family, while the average weight reflects the geometrical structure. Hence, it is reasonable to use them to estimate the cost or weight of the possible solutions.

9.3.3 *Exhaustive search*

In the alternative solution generating sub-module, two methods are provided to generate an optimal solution for a design input, namely, the exhaustive search and the GA-based search. For the online system that has been implemented, when a large volume of computation is involved in the user interaction activities, Web page requests will be terminated due to time-out on the users' side; and on the server side, it will also consume a large amount of computational resource and lower the efficiency of the entire system. Hence, the exhaustive search is used when the computational load is not very heavy, while the GA-based search will be employed to achieve the optimal results in an acceptable time when the exhausted search becomes too resource consuming.

The following steps are the procedures of the exhaustive search for an input function requirement string: '@M1010*C0010 @P1110 @P3110@E1110'.

Step 1. For each section in the function string, retrieve all the product families that satisfy the section according to the procedures presented in Section 9.3.2.2. For example, a number of $n1$ products are retrieved for @M1010, $n2$ for *C0010, etc.

Step 2. Next, all the possible solutions are generated. In this case, a total number of $n1*n2*n3*n4*n5$ solutions are generated.

Design Reuse for Embodiment and Detailed Design

Step 3. In all the possible solutions that have been generated, there might be some identical combinations. These identical combinations will be discarded to reduce the computational load.

Step 4. For the remaining solutions, the validity of each solution will be evaluated, and those invalid ones will be discarded.

Step 5. For the remaining solutions from Step 4, the cost or weight will be calculated for each solution.

Step 6. All the solutions will be ranked according to the optimization objective.

9.3.4 GA-based search

The computational load will become too high for a Web page application when the input function requirement string is too long or there are too many available product cases. In this book, genetic algorithm (GA) is adopted to search for an optimal solution for a given design problem these circumstances in order to ease the computational burden. The objective is to find a set of product families that can meet the desired function requirement, while at the same time minimize the cost or weight of the design.

9.3.4.1 Problem encoding

A chromosome embodies a potential solution for an optimal search problem, and is usually encoded as a binary or symbolic string to be handled computationally. In the GA-based search in this design reuse system, a chromosome represents a combination of component product families. Each gene represents one component product family that can fulfill one section in an input function requirement. Figure 9.4 illustrates the formation of a chromosome. For an input function requirement string with five function sections, a five-gene chromosome will be generated. Each gene is an index of a product family, which is a string beginning with a product family. The method to retrieve a product family for each gene has been described in Section 9.3.2.2. The validity of the chromosome can be evaluated using the process described in Section 9.3.2.3. For an n-section function requirement, the chromosome will have n genes.

Product Family Function:

PF203————@**M1010** @P1110 @E1010

PF435————@**C0010** @E3110

PF203————@M1010 @**P1110** @E1010

PF889————@**P3110***C0100 @M3000

PF405————@P1111 @**E1110**

Function Requirement String:

| @M1010 | *C0010 | @P1110 | @P3110 | @E1110 |

Possible Chromosome: PF203 - PF435 - PF203 - PF889 - PF405

Figure 9.4 Chromosome formation

9.3.4.2 Population Initialization

In a GA process, an initial generation, which is a group of valid chromosomes, has to be generated first. The population initialization procedure is described as follows:

Step 1. For each section of the function requirement, randomly choose a product family from the product database that can fulfill this section.

Step 2. Assign the index of the retrieved product family as the gene representing this function section in the chromosome.

Step 3. Repeat Steps 1 and 2 for each function section in the function requirement until a chromosome is formed.

Step 4. Repeat Step 3 to generate a prescribed number, which is the population size, of chromosomes.

9.3.4.3 Crossover operator

The crossover operator is used to create two new individuals (offsprings) from two existing individuals (parents) picked from the current population using a selection operation. There are several ways to achieve this such as a single-point crossover, two-point crossover, cycle crossover, and uniform crossover. An adapted uniform crossover is adopted in this book.

The popular uniform crossover [Ackley, 1987] works as follows: for each gene position 1 to L, randomly pick a gene from either of the two parent strings. This means that each gene is independent of the other genes. In the problem encoding addressed in this chapter, each gene is relatively independent of each other. Hence, for this problem, the uniform crossover will be a more effective operator to introduce diversity, overcome the limited information capacity of small populations and the tendency for more homogeneity. In this problem, the chromosomes represent tree-like product family combinations according to the input function requirement, which is also in the form of a tree-like data. The branch genes, which represent the main functions, are independent of each other. For the leaf genes, they are related to their branch genes. It is assumed that the leaf-branch relations in the function requirement are satisfied in the existing products. Hence, the uniform crossover happens at the branch level of the chromosomes. This means that a branch gene and its leaf genes are taken as a whole in the crossover to be inherited by the offspring. Figure 9.5 illustrates this process.

9.3.4.4 Mutation operator

In this procedure, all the individuals in the populations are checked gene by gene, and a certain proportion of the genes will be chosen to change. In the GA process, the mutation operator changes the value of one randomly selected gene.

Original Function Requirement

@M0010*C0000@P0200*C0
110*E1111@P2000@E0010

Function Requirement

M0010 P0200 P2000 E0010

C0000 C0110 E1111

Chromosome Crossover Process

Parent 1 PF203 – PF323 – PF12 – PF12 – PF101 – PF889 – PF405
Parent 2 PF401 – PF401 – PF105 – PF25 – PF36 – PF504 – PF338

Offspring 1 PF203 – PF323 – PF105 – PF25 – PF36 – **PF889** – PF338
Offspring 2 PF401 – PF401 – **PF12 – PF12 – PF101** – PF504 – **PF405**

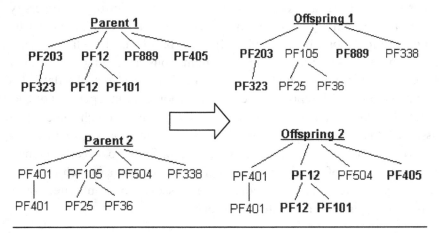

Figure 9.5 Adapted uniform crossover

9.3.4.5 Reproduction

The reproduction operator determines the way in which the strong individuals are reserved for reproduction and survival. Three commonly used selective strategies are the proportional selection, deterministic selection and ranking reproduction. The ranking strategy is adopted in this system. Equation (9.4) is used to calculate the selection probability of a chromosome. In this equation, p_i refers to the probability of a chromosome to be selected, of which its fitness ranks ith in the generation. The ranking is achieved by sorting all the chromosomes in the generation by their fitnesses in an ascending order. N is the population size of the generation.

$$p_i = \frac{2i}{N \times (N+1)} \tag{9.4}$$

9.3.4.6 Fitness function design

In this GA optimal search, a user can search based on two optimization goals, namely, minimum cost or minimum weight. The fitness for the two goals is calculated using Equations (9.5) and (9.6), respectively:

$$F_c = \frac{1}{\sum_{i=1}^{k} C_i} \tag{9.5}$$

$$F_w = \frac{1}{\sum_{i=1}^{k} W_i} \tag{9.6}$$

In Equation (9.5), F_c is the fitness of a chromosome in the search for minimum cost, C_i is the average cost of the product family that the ith gene represents, and k is the length of the chromosome. When a product family appears repetitively in different gene positions, its average cost will only be counted once.

Equation (9.6) is similar to Equation (9.5), except that all the values and symbols refer to the product weight instead of the product cost: F_w is the fitness of a chromosome in the search for minimum weight, W_i is the average weight of the product family that the ith gene represents, and k is the length of the chromosome. Similarly, when a product family appears repetitively in different gene positions, its average weight will only be counted once.

9.3.4.7 GA Iteration and configuration

In a GA process, after population initialization, an iteration of the crossover-mutation-reproduction operation will be conducted until certain stopping criteria have been met. The performance of the search is largely determined by the configuration of parameters, such as crossover rates, mutation rates and population size. A higher crossover rate with a lower mutation rate is recommended for most GA applications [Gen and Cheng 1997, Zalzala and Fleming 1997]. The population size has to be compromised between global optima and computational cost. In this system, the crossover and mutation rates are set at 0.7 and 0.01, respectively, and a population size of 100 was selected through a trial-and-error process to achieve the most efficient and stable GA process for the experimental data. A combinational stopping criterion is adopted: (1) when there is no further improvement of the fitness of the best chromosome in continuous five generations, the search is terminated, or (2) after an iteration of 300 generations, the search is terminated. When either criterion is met, the GA process will be stopped.

9.3.5 *Solution generation in washing machine design*

The alternative solution generating method is illustrated using an example of a two-way washing machine design. This product can function as a normal washing machine as well as a clothes dryer. Since there is no such product in the database, the users hope to retrieve several existing products to develop a new product using them as components.

In the first step (Figure 9.6), the users formalize the function requirement for the intended washing machine. The function requirement includes a few of the main functions of the major components. Through the interface in Figure 9.6, the users select the maximum cost as the optimization objective for this retrieval. Next, the system will retrieve and rank the suggested solutions based on their estimated costs.

Design Reuse for Embodiment and Detailed Design

In this process, the alternative solution generating sub-module searches the product database and suggests three solutions for the input function requirement. Figure 9.7 shows the result.

For each solution, several product families are retrieved and combined. The three solutions are listed with the first having the lowest estimated cost. All the member products are listed under the names of the retrieved product families. Users can check the information of the product families and their member products by clicking on their names to view their profile pages. Each component product family satisfies one or more entries of the input function. All the product families in the solutions can satisfy the entire input function requirement.

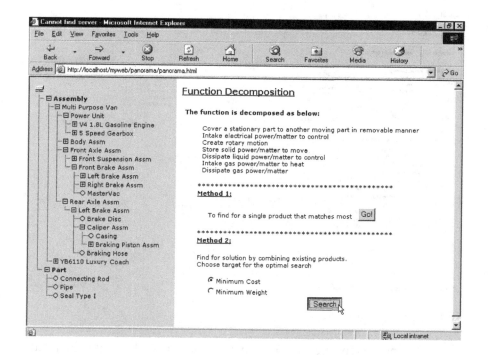

Figure 9.6 Interface: alternative solution generating sub-module

In the first solution, the system suggests combining a housing, a tumbler, a motor, a water controller and an air-heating pump to form a new product. The second solution achieves the function requirement by adding a water controller to an existing dryer machine. The third solution suggests adding an air-heating pump to an existing washing machine. Although the first solution has the lowest cost, it has many components. The possibility that the components are incompatible is also higher. Comparing the other two solutions, since the intended product is a washing machine, the technical feasibility of the third solution is higher than the second solution. In addition, the estimated cost is also within the budget cost. Thus, the third solution is adopted.

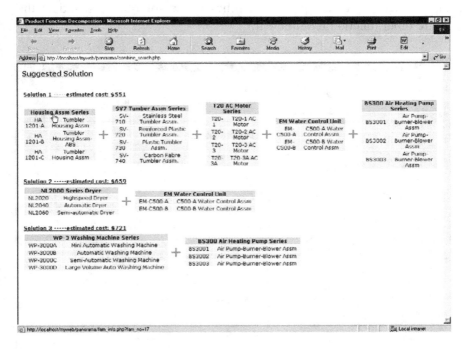

Figure 9.7 Generated solutions

In the next step, the idea to solve the design problem is generated. However, users still need to decide which products from the two product families are to be used. The users can input values for the desired product to find the most suitable product using the product retrieving sub-module. In this case, since the numbers of the member products and the KCs in the two product families are not too large, the users can easily compare all the products using the tables containing the KCs to make a decision.

Finally, the BS3001 air-heating pump and the WP3000-A washing machine are selected to form the new product. Users can refer to their product profile pages for more information and detailed design reuse. Through these four steps, the design problem is solved by combining two existing products.

9.4 Detailed Design

When the users have retrieved an appropriate existing design to reuse for a new design using information at either the concept level or the embodiment level, they still need more information of the geometrical details to make sure that the product is also reusable at the detailed design level. Hence, further processing is needed to generate a new detailed design based on the existing design. The detailed design transformation module completes this final step in the entire design reuse process to arrive at a new design with geometrical details.

9.4.1 *Architecture of the detailed design transformation*

This detailed design transformation module is based on the variant method and the concept of product family. It handles the interactions between the users and the CAD system. It is realized using two applications that communicate via Internet sockets (Figure 9.8). The first application is the HTTP (HyperText Markup Language) server program in an online server computer. The second application is a UG/Open API program supported by Unigraphics (UG) in an online UNIX host. This CAD system is not directly involved with the users. The HTTP application interacts with the users, recognizes the requests relating to the geometrical models, sends them to the UG/Open API program for further processing, and presents the returned information to the users for more interactions. The UG/Open API program handles the commands from the HTTP application, extracts the feature and parameter information from the CAD models, and modifies the existing CAD models, or generates new models with data from the

264 *Design Reuse in Product Development Modeling, Analysis and Optimization*

users. In this way, the system does not require the users to be skillful in using CAD tools, and allows the users to manipulate 3D CAD models through a general online interface.

9.4.1.1 HTTP server program

The HTTP server application has three main functions, namely product data browsing, product searching and retrieving, and communication with the UG/API server program. These three functions are briefly described next.

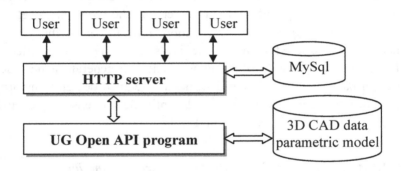

Figure 9.8 Implementation architecture for detailed design transformation

Product data browsing. Users can browse the hierarchical structure of the design database. At the same time, the users can also view the detailed description of any specific design with a graphical display.

Product searching and retrieving. The server application accepts the design input from the users, such as the functional requirements or other search criteria, and searches the design database to retrieve the most approximate existing design to reuse.

Communication with the UG/API server program. Users' requests relating to geometrical data processing will be sent to the UG/API server program via the Internet socket communication. Next, the HTTP server application will receive feedback from the UG/API server program to answer the users' requests.

9.4.1.2 UG/API server program

The two main functions of the UG/API server program are detailed design reuse and graphical file generation.

Detailed design reuse. The UG/API programs are capable of generating new detailed designs from existing design data based on the variant method.

Graphical file generation. Currently, due to the constraint of the Internet bandwidth, it is still not viable to display original CAD data files, such as UG part files in this system via the Internet. In this system, the VRML file format is adopted for online graphical display. A VRML plug-in can be embedded in a Web browser to facilitate visualization of the geometric models. Each time a user browses certain design data or tries to reuse it, the UG/API server program will instantly generate an up-to-date VRML format file of the design and send it to the HTTP server for display.

9.4.1.3 Design descriptions and data structure

In the present online Web-based system, the description of a specific product design consists of two parts, namely the product attributes and a geometrical model. The product attributes include functions, KCs, cost, etc. They are all text-based, and are stored in a relational database. As for the geometrical models, all the 3D models are stored in the UG part file format and reside on an online UNIX host.

9.4.2 *Feature-based parametric modeling*

This is the most widely used modeling method in mainstream solid modelers. It is a combination of the feature-based and parametric-based modeling methods [Stork *et al.*, 1996; Monedero, 2000; Brunetti and

Golob, 2000]. In feature-based modeling, a solid model is constructed by combining features, such as blocks, cylinders, holes, chamfers, etc. Parametric-based modeling offers the ability to parameterize various geometrical aspects of the design models, generally to set various dimensional elements of the model geometry, such as the length, position, diameter, etc. These parameters can be assigned physical, numerical values and can be altered at will. The feature-based parametric modeling approach modularizes the geometrical modeling and makes it easier and more intuitive to make use of existing 3D models and modify them. For example, to generate a new detailed design based on some existing designs, a user simply needs to decide the features to add or remove and the parameters to alter. The solid modeler will then generate a solid model for the new design [Andrews and Sivaloganathan, 1998; Andrews *et al.*, 1999; Finger and Safier, 1990; Chung *et al.*, 1990].

However, the feature-based parametric modeling approach is not directly implemented in the system as described in this chapter. It is combined with the concept of product family to facilitate further design reuse and data management.

9.4.3 *Product family and variant method*

The concept of a product family has different meanings and is implemented differently at different product life-cycle phases. In this system, products that share the same functions and have similar geometrical structures are clustered into a product family. This is to simplify product data management, facilitate data retrieval and reuse, and also allows the concept to be easily implemented with the UG/Open API.

A product family can be seen as a virtual spreadsheet that records the information of a group of products. It consists of two parts, namely the family definition and the member data. The family definition is made up of a series of characteristic features and parameters of the products that are mostly of concern to the designers. The member data include the ID of the member products and the values of their attributes for the family definition. The value for a feature attribute is either 'yes' or 'no', indicating whether the feature is included or excluded in the member product. For the parameters, the values are the desired numerical values for the individual member products.

In the Web-based system, in order to reuse an existing design, a user needs to provide all of the desired values for all the family attributes, and a new member product will be added to the family. If the user wishes to control more features and parameters, he/she can edit the family definition by adding in more attributes. This is the variant approach to generate a new design. In the variant approach, there is usually a 'master model' that contains all the combined features of a family of designs. The master model is used to build an instance of an individual family member [Andrews and Sivaloganathan, 1998].

9.4.4 *Operation of detailed design reuse*

9.4.4.1 Product data browsing

Since the online Web-based environment is based on dynamic Web pages, it takes advantage of the cross-referencing feature of hypertext markup language (HTML). In the information page of each product, related products within the product hierarchy or the peer products in the same product family are listed, and the users can access their profiles by simply clicking on their name. Figure 9.9 shows a product profile interface.

9.4.4.2 Design transformation

This is the core of the detailed design reuse process. At this step, a new design is generated and the user's design intent is realized. As mentioned earlier, the design transformation is based on the concept of product family, which means that a new design is achieved through manipulations of an existing product family. All the available manipulations are addressed next, and the design transformation process is self-evident.

Product family creation. When a satisfactory product has been retrieved for reuse based on a user's requirements, the user needs to check whether this product belongs to any existing product families. If it does, the user can generate a new design by adding a member into this family; otherwise, the user needs to first define a new product family, that is, to specify the attributes in a product family definition. This checking mechanism is performed through a set of similarity metrics discussed in Section 9.3.1. At the user's request, all the features and parameters in the solid model of the retrieved product will be extracted and presented to the user as a dynamic HTML template. The user can choose those features and attributes that may vary for each different member product, and the selections will be sent back to the UG/API server program for processing. As such, a new family definition will be created.

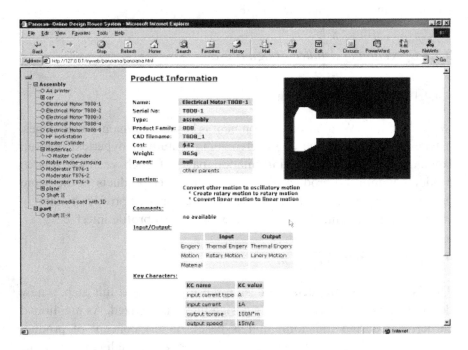

Figure 9.9 Product profile interface

Adding member product. When a product family is available, the user will be prompted with a menu to provide all the desired values for the new design, as shown in Figure 9.10. All these data will be submitted to the UG/API program, and the solid model for the new design is generated based on the variant method. The new design will be stored in the database. The user can instantly view the result from the graphical display on the Web browser.

Editing product family definition. This function allows the user to modify some features or parameters of the product family. For example, more features or parameters can be added to the product family definition (Figure 9.11).

Altering existing design data. An existing detailed design can also be modified in this online Web-based environment. For a certain product, all the family attributes and their corresponding values will be extracted; the user can make alterations in the interface shown in Figure 9.12 and submit them. The solid model of the product will be modified correspondingly.

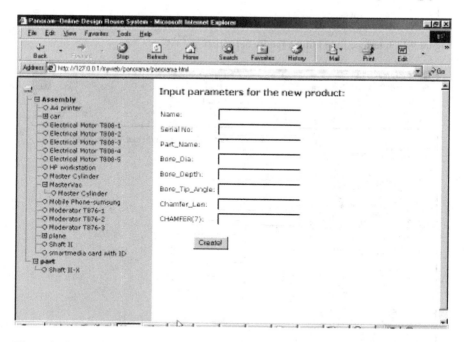

Figure 9.10 Interface for adding a new member product

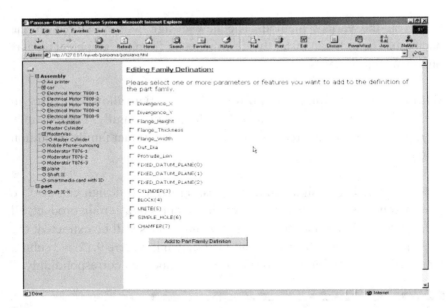

Figure 9.11 Interface for editing a family definition

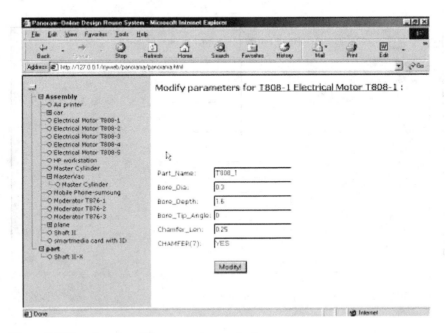

Figure 9.12 Interface for editing a member product

9.4.5 *System implementation*

In the online Web-based environment, the HTTP server application is developed using Hypertext Preprocessor (PHP), and the UG/API server application is developed using the C language based on the UG/Open API on an online UNIX host. The two applications communicate via the Internet socket. MySql is the adopted relational database. All experimental geometrical data are created with the UG solid modeler V15. An Apache server is used for Web publishing. In the prototyping stage, a Blaxxun 3D viewer is used as the embedded viewer of Internet Explorer for the VRML files.

9.5 Summary

Embodiment and detailed design are necessary to develop the product concepts into real products. With an emphasis on the computational support to the processes, this chapter proposes an online Web-based design reuse system to provide a distributed, efficient and low-cost means for embodiment design and detailed design reuse.

In the embodiment design stage, the chapter focuses on the generation of candidate solutions based on various case retrieval and optimization methods. In particular, the system features a function decomposition process and a subsequent search process for solutions that fulfill the functional requirements. A GA-based search method is developed to achieve rapid generation and evaluation of the candidate solutions.

The detailed design transformation application is used for detailed design. It reduces the common difficulties faced by users in handling sophisticated CAD systems in detailed design reuse, and makes the existing product data more meaningful and accessible. Another advantage of this online Web-based environment is that for many small and medium enterprises, their products usually fall into several product families, and many of their design tasks can be achieved through modifications of these existing designs. Hence, with this online Web-based system, the resource of the CAD system can be shared throughout the enterprises.

Bibliography

Ackley, D. (1987). A Connectionist Machine for Genetic Hillclimbing, Boston: Kluwer.
Adriaans, P. and Zantinge, D. (1996). Data Mining, Harlow: Addison-Wesley
Allen, K.R. and Carlson-Skalak, S. (1998). Defining Product Architecture During Conceptual Design, ASME Design Engineering Technical Conferences - Design Theory & Methodology Conference, Paper No. DETC98/DTM-5650, Atlanta, GA, September.
Allenby, B.R. (1991). Design for Environment: A Tool Whose Time Has Come, SSA Journal, **12**(9), pp. 5–9.
Alting, L. and Legarth, J.B. (1995), Life Cycle Engineering and Design, Annals for the CIRP, **44**(2), pp. 569–580.
Altmeyer, J., Ohnsorge, S. and Schurmann, B. (1994). Reuse of Design Object in CAD Frameworks, In: Proceedings of the IEEE International Conference on Computer Aided Design (ICCAD), San Jose, CA, USA, November.
Altmeyer, J. and Schurmann, B. (1996). On Design Formalization and Retrieval of Reuse Candidates, Artificial Intelligence in Design'96, pp. 231–250.
Altshuller G.S. (1984). Creativity as an Exact Science, New York: Gordon and Breach.
Ambler, S.W. (2004). The Object Primer: Agile Model Driven Development with UML 2, New York: Cambridge University Press.
Andrews, P.T.J. and Sivaloganathan, S. (1998). A Variant Model for Storing Families of Mechanical Designs, In: Design Reuse, Engineering Design Conference' 98, Sivaloganathan, S. and Shahin, T.M.M., (eds), 1999, London: Professional Engineering Publishing, pp. 361–370.
Andrews, P.T.J., Shahin, T.M.M. and Sivaloganathan, S. (1999). Design Reuse in a CAD Environment – Four Case Studies, Computers and Industrial Engineering, **37**, pp. 105–109.
Andersson, J. and Wallace, D. (2002). Pareto Optimization Using the Struggle Genetic Crowding Algorithm, Engineering Optimization, **34**, pp. 623–644.
Avriel, M. (1976). Nonlinear Programming: Analysis and Methods, Englewood Cliffs, N.J.: Prentice-Hall.

Bahrami, A. (1994). Routine Design with Information Content and Fuzzy Quality Function Deployment, Journal of Intelligent Manufacturing, **5**(4), pp. 203–210.

Baldwin, C.Y. and Clark, K.B. (2000), Design Rules: Volume 1, The Power of Modularity, Cambridge, MA.: MIT Press.

Baxter, J., Juster, N. and de Pennington, A. (1994). A Functional Data Model for Assemblies Used to Verify Product Design Specifications, In: Proceedings of the Institution for Mechanical Engineers, Part B – Journal of Engineering Manufacture, **208**, pp. 235–244.

Bergmann, R. (2002). Experience Management: Foundations, Development Methodology, and Internet-based Applications, New York: Springer.

Berry, M.J.A. and Linoff, G. (1997). Data Mining Techniques: For Marketing, Sales, and Customer Support, New York: John Wiley.

Bilgic, T. and Rock, D. (1997). Product Data Management Systems: State-of-the-Art and Future, In: Proceedings of 1997 ASME Design Engineering Technical Conferences and Computers in Engineering Conference, Paper No. DETC97/EIM-3720, Sacramento, CA, September.

Blackenfelt, M. and Sellgren, U. (2000). Design of Robust Interfaces in Modular Products, ASME Design Engineering Technical Conferences- Design Automation Conference, Paper No. DETC00/DAC-14486, Baltimore, MD., September.

Bobrow, D., Falkenhainer, B., Farquhar, A., Fikes, R., Forbus, K.D., Gruber, T.R., Iwasaki, Y. and Kuipers, B.J. (1996). A Compositional Modeling Language, In: Proceedings of the 10th International Workshop on Qualitative Reasoning, Menlo Park, CA., May, AAAI Press, pp. 12–21.

Brunetti,G. and Golob, B. (2000). A Feature-Based Approach Towards a Integrated Product Model Including Conceptual Design Information, Computer-Aided Design, **32**(14), pp. 877–887

Burgess, J.D.G., Shahin, T.M.M. and Sivaloganathan, S. (1998). Design reuse for Detailed Component Design, Case Study: Development of Component Design Reuse System for NiftyLift PLC, In: Design Reuse, Engineering Design Conference' 98, Sivaloganathan, S. and Shahin, T.M.M., (eds), 1999, London: Professional Engineering Publishing, pp. 271–282.

Busby, J.S. (1999). The Problem with Design Reuse: An Investigation into Outcomes and Antecedents, Journal of Engineering Design, **10**(3), pp. 277–296.

Campbell, M., Cagan, J. and Kotovsky, K. (1999). A-Design: an Agent-based Approach to Conceptual Design in a Dynamic Environment, Research in Engineering Design, **11**, pp. 172–192.

Campbell, M., Cagan, J., Kotovsky, K. (2000). Agent-based Synthesis of Electromechanical Design Configurations, Journal of Mechanical Design, **122**, pp. 61–69.

Chakrabarti, A. and Bligh, T. (1996). An Approach to Functional Synthesis of Mechanical Design Concepts: Theory, Applications, and Emerging Research Issues, Artificial Intelligence for Engineering Design, Analysis and Manufacturing, **10**, pp. 313–331.

Chakrabarti, A. (2002). Engineering Design Synthesis, Understanding, Approaches and Tools, New York: Springer.

Chandrasekaran, B., Goel, A. and Iwasaki, Y. (1993). Functional Representation as Design Rationale, IEEE Computer, 26(1), pp. 48–56.

Charnes, A. and Cooper, W.W. (1977). Goal Programming and Multiple Objective Optimization (Part 1), European Journal of Operational Research, 1(1), pp. 39–54.

Chen, A., McGinnis, B., Ullman, D. and Dietterich, T. (1990). Design History Knowledge Representation and Its Basic Computer Implementation, In: The 2nd International Conference on Design Theory and Methodology, ASME, Chicago, IL, pp. 175–185.

Chen, W., Allen J.K., Mavris, D.N. and Mistree, F. (1996). A Concept Exploration Method for Determining Robust Top-Level Specifications, Engineering Optimization, 26, pp. 137–158.

Chidambaram, B. and Agogino, A.M. (1999). Catalog-based Customization, In: Proceedings of 1999 ASME Design Engineering Technical Conferences- Design Automation Conference, Paper No. DETC99/DAC-8675, Las Vegas, Nevada, September.

Chung, J., Jack, C.H. and Schussel, M.D. (1990). Technical Evaluation of Variational and Parametric Design, In: Proceedings of ASME International Computers in Engineering Conference and Exhibition, Boston, MA, USA, 4–8 August, pp. 289–298.

Clark, P. and Matwin, S. (1993). Using Qualitative Models to Guide Inductive Learning, In: Proceedings of the 1993 International Conference on Machine Learning, CA. Kaufmann, pp. 49–56.

Clarkson, P.J. (1998). The Cambridge Engineering Design Center, In: Design Reuse, Engineering Design Conference' 98, Sivaloganathan, S. and Shahin, T.M.M., (eds), 1999, London: Professional Engineering Publishing, pp.103–112.

Clausing, D. (1994). Total Quality Development: a Step-by-Step Guide to World Class Concurrent Engineering, New York: ASME Press.

Clausing, D. (1998). Reusability in Product Development, In: Design Reuse, Engineering Design Conference' 98, Sivaloganathan, S. and Shahin, T.M.M., (eds), 1999, London: Professional Engineering Publishing, pp. 57–66.

Cohen, L. (1995). Quality Function Deployment: How to Make QFD Work for You, Reading, Mass.: Addison-Wesley.

Collier, D.A. (1981). The Measurement and Operating Benefits of Component Part Commonality, Decision Sciences, 12(1), pp. 85–96.

Conklin, J. and Burgess-Yakamovic, K. (1991). A Process-Oriented Approach to Design Rationale, Human-Computer Interaction, 6(3,4), pp. 357–391.

Cooper, R. and Kaplan, R.S. (1991). The Design of Cost Management Systems, Englewood Cliffs, NJ.: Prentice-Hall.

Corbett, B. and Rosen, D.W. (2004). A Configuration Design Based Method for Platform Commonization for Product Families, Artificial Intelligence for Engineering Design, Analysis and Manufacturing, 18, pp. 21–39.

Counsell, J., Porter, I., Dawson, D. and Duffy, M. (1999). Schemebuilder: Computer Aided Knowledge Based Design of Mechatronic Systems, Assembly and Automation, **19**(2), pp. 129–138.

Dahmus, J.B., Gongzalez-Zugasti, J.P. and Otto, K.N. (2000). Modular Product Architecture, In: Proceedings of the 2000 ASME Design Theory and Methodology Conference, Paper No. DETC2000/DTM-14565, Baltimore, MD, September.

Dai, Z. and Scott, M.J. (2004). Effective Product Family Design using Preference Aggregation, In: ASME Design Engineering Technical Conference – Design Theory and Methodology Conference, Paper No. DETC2004/DTM-57419, Salt Lake City, UT. September.

Darr, T.P. and Birmingham, W.P. (1994). Automated Design for Concurrent Engineering, IEEE Expert: Intelligent Systems and Their Applications, **9**(5), pp. 35–42.

Deb, K. (2001). Multi-Objective Optimization Using Evolutionary Algorithms, New York: John Wiley & Sons.

Delionback, L.M. (1995). Learning Curves and Progress Functions, In: Cost Estimator's Reference Manual, 2nd ed., Stewart, R., Wyskida, R. and Johannes, J., (eds.), New York: John Wiley & Sons, 1995, pp. 169–192.

de Weck, O., Suh, E.S. and Chang, D. (2003). Product Family and Platform Portfolio Optimization, In: Proceedings of ASME 2003 Design Engineering Technical Conferences – Design Automation Conference, Paper No. DETC2003/DAC-48721, Chicago, Illinois, September.

Dong, Q. and Whitney, D.E. (2001). Designing a Requirement Driven Product Development Process, In: Proceedings of ASME 2001 Design Engineering Technical Conferences and Computers and Information in Engineering Conferences, Paper No. DETC2001/DTM-21682, Pittsburgh, Pennsylvania, September.

D'Souza, B. and Simpson, T.W. (2003). A Genetic Algorithm Based Method for Product Family Design Optimization, Engineering Optimization, **35**(1), pp. 1–18.

Du, X., Jiao, J. and Tseng, M.M. (2001). Architecture of Product Family: Fundamentals and Methodology, Concurrent Engineering: Research and Applications, **9**(4), pp. 309–325.

Du, X., Jiao, J. and Tseng, M.M. (2002a), Graph Grammar Based Product Family Modelling, Concurrent Engineering: Research and Applications, **10**(2), pp. 113–128.

Du, X., Jiao, J. and Tseng, M.M. (2002b), Product Family Modeling and Design Support: An Approach Based on Graph Rewriting Systems, Artificial Intelligence for Engineering Design, Analysis and Manufacturing, **16**(2), pp. 103–120.

Duffy, A.H.B. and Duffy, S.M. (1996). Learning for Design Reuse, Artificial Intelligence for Engineering Design, Analysis and Manufacturing, **10**, pp. 139–142.

Duffy, A.H.B. and Ferns, A.F. (1999). An Analysis of Design Reuse Benefits, In: Proceedings of the ICED 99 Conference, Lindemann, U., Birkhofer, H., Meerkamm, H. and Vajna, S., (eds.), Technische Universität München, 1999, pp. 799–804.

Duffy, A.H.B. and Kerr, S.M. (1993). Customized Perspectives of Past Designs from Automated Group Rationalizations, Artificial Intelligence in Engineering, **8**, pp. 183–200.

Duffy, S.M., Duffy, A.H.B. and MacCallum, K.J. (1995). A Design Reuse Process Model, In: Proceedings of the International Conference on Engineering Design (ICED95), Prague, August, Heurista Zurich, pp. 490–495.

Duffy, A.H.B., Persidis, A. and MacCallum, K.J. (1996). NODES: A Numerical and Object Based Modeling System for Conceptual Engineering Design, Knowledge-Based Systems, **9**, pp. 183–206.

Duffy, A.H.B., Smith, J.S. and Duffy, S.M. (1998). Design Reuse Research: a Computational Perspective, In: Design Reuse, Engineering Design Conference' 98, Sivaloganathan, S. and Shahin, T.M.M., (eds), 1999, London: Professional Engineering Publishing, pp. 43–56.

Ehrlenspiel, K. (1985). Kostengünstig Konstruieren (Cost-Effective Designing), Berlin and New York: Springer-Verlag.

Eppinger, S.D., Whitney, D.E., Smith, R.P. and Gebala, D.A. (1994). A Model-Based Method for Organizing Tasks in Product Development, Research in Engineering Design, **6**(1), pp. 1–13.

Erens, F.J., McKay, A. and Bloor, S. (1994). Product Modeling Using Multiple Levels of Abstraction Instances as Types, Computers in Industry, **24**(1), pp. 17–28.

Ericsson, A. and Erixon, G. (1999). Controlling Design Variants: Modular Product Platforms, New York: ASME Press.

Erixon, G. (1996). Design for Modularity, In: Design for X – Concurrent Engineering Imperatives, Huang, G.Q. (ed.), pp. 356–379, New York: Chapman & Hall.

Erixon, G. and Ostgren, B.M. (1993). Synthesis and Evaluation Tool for Modular Design, International Conference on Engineering Design (ICDE '93), Hague, Netherlands, August, pp. 898–905.

Erixon, G., von Yxkull, A. and Arnström, A. (1996). Modularity - the Basis for Product and Factory Reengineering, Annals of the CIRP, **45**(1), pp. 1–4.

Erlandsson, A., Erixon, G. and Ostgren, B. (1992). Product Modules – the Link Between QFD and DFA?, The International Forum on Product Design for Manufacture and Assembly, Newport, RI., June.

Faltings, B. (2002). FAMING: Supporting Innovative Design Using Adaptation – a Description of the Approach, Implementation, Illustrative Example and Evaluation, In: Engineering Design Synthesis, Understanding, Approaches and Tools, Chakrabarti, A., (ed.), New York: Springer, pp. 285–302.

Feitzinger, E. and Lee, H.L. (1997). Mass Customization at Hewlett-Packard: The Power of Postponement, Harvard Business Review, **75**(1), pp. 116–121.

Felfernig, A., Friedrich, G. and Jannach, D. (2001). Conceptual Modeling for Configuration of Mass-Customizable Products, Artificial Intelligence in Engineering, **15**(2), pp. 165–176.

Fellini, R., Papalambros, P. and Weber, T. (2000). Application of Product Platform Design Process to Automotive Powertrains, In: 8th AIAA/NASA/USAF/ISSMO Symposium on Multidisciplinary Analysis and Optimization, Long Beach, CA, AIAA, Paper No. AIAA-2000-4849.

Fensel, D., Erdmann M. and Studer, R. (1997). Ontology-groups: Semantically Enriched Subnets of the WWW, In: Proceedings of the 1st International Workshop Intelligent Information Integration during the 21st German Annual Conference on Artificial Intelligence, Freiburg, Germany, September.

Finger, S. (1998) Design Reuse and Design Research, In: Design Reuse, Engineering Design Conference' 98, Sivaloganathan, S. and Shahin, T.M.M., (eds), 1999, London: Professional Engineering Publishing, pp.3–9.

Fisher, M., Ramdas, K. and Ulrich, K. (1999). Component Sharing in the Management of Product variety: a Study of Automotive Braking Systems, Management Science, **45**(3), pp. 297–315.

Finger, S. and Safier, S.A. (1990). Representing and Recognizing Features in Mechanical Designs, In: SME Design Theory and Methodology Conference, **27**, pp. 19–25.

Fixson, S.K. (2002). Linking Modularity and Cost: A Methodology to Assess Cost Implications of Product Architecture Differences to Support Product Design, Ph.D. Thesis, Massachusetts Institute of Technology, Cambridge, MA.

Fonseca, C. and Fleming, P. (1998). Multiobjective Optimization and Multiple Constraint Handling with Evolutionary Algorithms – Part I: A Unified Formulation, IEEE Transactions on Systems, Man, and Cybernetics – Part A: Systems & Humans, **28**(1), pp. 26–37.

Fourman, M.P. (1985). Compaction of Symbolic Layout Using Genetic Algorithms, In: Genetic Algorithms and Their Applications: Proceedings of 1st International Conference on Genetic Algorithms, Grefenstette, J.J., (ed.), Princeton, NJ: Lawrence Erlbaum, pp. 141–153.

Fowlkes, W.Y. and Creveling, C.M. (1995). Engineering Methods for Robust Product Design Using Taguchi Methods in Technology and Product Development, Reading, Mass.: Addison-Wesley.

Fujita, K. (2002). Product Variety Optimization under Modular Structure, Computer-Aided Design, **34**, pp. 953–965.

Fujita, K., Akagi, S., Yoneda, T. and Ishikawa, M. (1998a). Simultaneous Optimization of Product Family Sharing System Structure and Configuration, In: Proceedings of the 1998 ASME Design Engineering Technical Conferences, Paper No. DETC98/DFM-5722, Atlanta, Georgia, September.

Fujita, K., Hirokawa, N. and Akagi, S. (1998b). Multi-Objective Optimal Design of Automobile Engine Using Genetic Algorithm, In: Proceedings of ASME 1998 Design Engineering Technical Conferences, Paper No. DETC1998/DAC-5799, Atlanta, Georgia, September.

Fujita, K. and Yoshida, H. (2001). Product Variety Optimization: Simultaneous Optimization of Module Combination and Module Attributes, In: Proceedings of

ASME 2001 Design Engineering Technical Conferences and Computers and Information in Engineering Conferences, Paper No. DETC2001/DAC-21058, Pittsburgh, Penn., September.

Fujita, K. and Yoshioka, S. (2003). Optimal Design Methodology of Common Components for a Class of Products: Its Foundations and Promise, In: Proceedings of ASME 2003 Design Engineering Technical Conferences and Computers and Information in Engineering Conferences, Paper No. DETC2003/DAC-48718, Chicago, Illinois, September.

Fujita, K., Sakaguchi, H., Akagi, S. and Yoneda, T. (1999). Product Variety Development and its Optimization under Modular Architecture and Module Commonalization, In: Proceedings of the 1999 ASME Design Engineering Technical Conferences, Paper No. DETC99/DFM-8923, Las Vegas, Nevada, September.

Gen, M. and Cheng, R. (1997). Genetic Algorithms and Engineering Design, New York: Wiley.

Gero, J.S. (1990). Design Prototypes: A Knowledge Representation Schema for Design, AI Magazine, 11(4), pp. 26–36.

Girczyc, E. and Carlson, S. (1993). Increasing Design Quality and Engineering Productivity through Design Reuse, In: Proceedings of the 30th International Conference on Design Automation, Dallas, Texas, US., pp: 48–53

Göker, M.H. and Roth-Berghofer, T. (1999). The Development and Utilization of the Case-based Help-desk Support System HOMER, Engineering Applications of Artificial intelligence, 12, pp. 665–680.

Goldberg, D., 1989, Genetic Algorithms in Search and Machine Learning, Reading, Mass.: Addison-Wesley.

Gonzalez-Zugasti, J.P. and Otto, K.N. (2000). Modular Platform-based Product Family Design, In: Proceedings of the 2000 ASME Design Engineering Technical Conferences and Computers and Information in Engineering Conference, Paper No. DETC2000/DAC-14238, Baltimore, Maryland, September.

Gonzalez-Zugasti, J.P., Otto, K.N. and Baker, J.D. (2000). A Method for Architecturing Product Platforms, Research in Engineering Design, 12, pp. 61–72.

Gonzalez-Zugasti, J.P., Otto, K.N. and Baker, J.D. (2001). Assessing Value in Platformed Product Family Design, Research in Engineering Design, 13, pp. 30–41.

Gorti, S.R. and Sriram, R.D. (1996). From Symbol to Form: A Framework for Conceptual Design, Computer-Aided Design, 28(11), pp. 853–870.

Grante, C. and Andersson, J. (2003). A Method for Evaluating Functional Content in Mechatronic Systems, Research in Engineering Design, 14, pp. 224–235.

Griss, M.L. (1991). Software Reuse at Hewlett-Packard, in Proceedings of the 1st International Workshop on Software Reusability, Dortmund, Germany, July, pp. 18–24.

Grueninger, T. and Wallace, D. (1996). Multi-modal Optimization Using Genetic Algorithms, Technical Report 96.02, CADlab, Massachusetts Institute of Technology, Cambridge.

Gu, P. and Sosale, S. (1999). Product Modularization for Life Cycle Engineering, Robotics and Computer Integrated Manufacturing, **15**, pp. 387–401.

Harry, M.J. and Schroeder, R. (2000). Six Sigma: the Breakthrough Management Strategy Revolutionizing the World's Top Corporations, New York: Doubleday.

Hata, T., Kimura, F. and Suzuki, H. (1997). Product Life Cycle Design Based on Deterioration Simulation, In: Life Cycle Networks, 4th CIRP International Seminar on Life Cycle Engineering, Krause, F.-L. and Seliger, G., (eds.), London: Chapman & Hall, pp. 59–68.

Haykin, S. (1999). Neural Networks: A Comprehensive Foundation, 2nd ed., Upper Saddle River, NJ: Prentice Hall.

Hernandez, G., Simpson, T. W., Allen, J. K., Bascaran, E., Avila, L. F. and Salinas, F. (2001). Robust Design of Families of Products with Production Modeling and Evaluation, ASME Journal of Mechanical Design, **123**(2), pp. 183–190.

Hernandez, G., Allen, J.K. and Mistree, F. (2002). Design of Hierarchic Platforms for Customizable Products, In: Proceedings of the ASME Design Engineering Technical Conferences - Design Automation Conference, Fadel, G., (ed.), Montreal, Quebec, Canada, ASME, Paper No. DETC2002/DAC-34095.

Hernandez, G., Allen, J. K. and Mistree, F. (2003). Platform Design for Customizable Products as a Problem of Access in a Geometric Space, Engineering Optimization, **35**(3), 229–254.

Hirtz, J., Stone, R, McAdams, D., Szykman, S. and Wood, K. (2002). A Functional Basis for Engineering Design: Reconciling and Evolving Previous Efforts, Research in Engineering Design, **13**, pp. 65–82.

Hommes, Q.D.V.E. and Whitney, D.E. (2003). The Predictability of System Interactions at Early Phase of the Product Development Process, In: Proceedings of ASME 2003 Design Engineering Technical Conferences and Computers and Information in Engineering Conferences, Paper No. DETC2003/DTM-48635, Chicago, Illinois, September

Hopp, W.J. and Spearman, M.L. (2001). Factory Physics, Boston, MA.: McGraw-Hill.

Hölttä-Otto, K. (2005). Modular Product Platform Design, Ph.D. Dissertation, Department of Mechanical Engineering, Helsinki University of Technology, Finland.

Hölttä, K. and Salonen, M. (2003). Comparing Three Modularity Methods, In: Proceedings of ASME 2003 Design Engineering Technical Conferences and Computers and Information in Engineering Conferences, Paper No. DETC2003/DTM-48649, Chicago, Illinois, September

Hölttä, K., Tang, V. and Seering, W.P. (2003). Modularizing Product Architectures Using Dendrograms, In: Proceedings of the International Conference on Engineering Design (ICED03), Stockholm, August.

Huang, C.C. and Kusiak, A. (1998). Modularity in Design of Products and Systems, IEEE Transactions on Systems, Man, and Cybernetics – Part A: Systems & Humans, **28**(1), pp. 66–77.

Huang, G.Q., Zhang, X.Y. and Liang, L. (2005). Towards Integrated Optimal Configuration of Platform Products, Manufacturing Processes, and Supply Chains, Journal of Operations Management, **23**(3-4), pp. 267–290.

Hundal, M.S. (1990). A Systematic Method for Developing Function Structures, Solutions and Concept Variants, Mechanism and Machine Theory, **25**(3), pp. 243–256.

Hundal, M.S. (1997). Systematic Mechanical Designing: a Cost and Management Perspective, New York: ASME Press.

Ignizio, J.P., 1982, Linear Programming in Single- & Multiple-Objective Systems, Englewood Cliffs, NJ: Prentice-Hall.

Ignizio, J.P. and Cavalier, T.M. (1994). Linear Programming, London: Prentice-Hall.

International Organization for Standardization. (1994a). ISO 10303-I:1994 – Industrial Automation Systems and Integration – Product Data Representation and Exchange – Part I: Overview and Fundamental Principles.

International Organization for Standardization. (1994b). ISO 10303-II:1994 – Industrial Automation Systems and Integration – Product Data Representation and Exchangep – Part II: the EXPRESS Language Reference Manual.

Ishibuchi, H., Yamamoto, N., Murata, T. and Tanaka, H. (1994). Genetic Algorithms and Neighborhood Search Algorithms for Fuzzy Flowshop Scheduling Problems, Fuzzy Sets and Systems, **67**(1), 81–100.

Iwasaki, Y. and Chandrasekaran, B. (1992). Design Verification Through Function- and Behavior-Oriented Representation, In: Proceedings of the Conference on Artificial Intelligence in Design '92, Gero, J.S., (ed.), Kluwer Academic Publishers, pp. 597–616.

Iwasaki, Y., Vescovi, M, Fikes, R. and Chandrasekaran, B. (1995). Causal Functional Representation Language with Behavior-Based Semantics, Applied Artificial Intelligence, **9**, pp. 5–31.

Jennings, N.R. (1995). Controlling Cooperative Problem Solving in Industrial Multi-agent Systems Using Joint Intentions, Artificial Intelligence, **75**(2), pp. 195–240.

Jensen, T. (1998). A Taxonomy for Design Reuse Systems: Proposing a System for Formalized Online Knowledge Capturing, In: Proceedings of the Engineering Design Conference, Brunel University, 23–25 June 1998, pp. 483–494

Jette, C. and Smith, R. (1989). Examples of Reusability in an Object-Oriented Programming Environment, in Software Reusability, Volume II, Applications and Experience, Biggerstaff, T.J. and Perlis, A.J. (eds), New York: ACM Press, Reading, Mass.: Addison-Wesley., pp. 73–101.

Jiang, L. and Allada, V. (2001). Design for Robustness of Modular Product Families for Current and Future Markets, In: Proceedings of ASME Design Engineering Technical Conferences – Design for Manufacturing Conference, Pittsburgh, PA., September, Paper No. DETC2001/DFM-21177.

Jiao, J. and Tseng, M.M. (1998). Fuzzy Ranking for Conceptual Evaluation in Configuration Design for Mass Customization, Concurrent Engineering: Research and Applications, **6**(3), pp. 189–206.

Jiao, J. and Tseng, M.M. (1999). An Information Modeling Framework for Product Families to Support Mass Customization Production, Annals of the CIRP, **48**(1), pp. 93–98.

Jiao, J. and Tseng, M.M. (2000). Understanding Product Family for Mass Customization by Developing Commonality Indices, Journal of Engineering Design, **11**(3), pp. 225–243.

Jiao, J. and Tseng, M.M. (2004). Customizability Analysis in Design for Mass Customization, Computer-Aided Design, **36**, pp. 745–757.

Jiao, J. and Zhang, Y. (2005). Product Portfolio Identification Based on Association Rule Mining, Computer-Aided Design, **37**, pp. 149–172.

Jiao, J., Pokharel, S., Zhang, L. and Zhang Y. (2005). Coordination of Product and Process Variety in Mass Customization with Data Mining Approach, In: 10th International Conference on Industrial Engineering Theory, Applications and Practice, Clearwater Beach, FL., December.

Johnson, M.R. and Wang, M.H. (1998). Economical Evaluation of Disassembly Operations for Recycling, Remanufacturing and Reuse, International Journal of Production Research, **36**(12), 3227–3252.

Jovane, F., Alting, L., Armillotta, A., Eversheim, W., Feldmann, K., Seliger, G. and Roth, N. (1993). A Key Issue in Product Life Cycle: Disassembly. Annals of the CIRP, **42**(2), 651–658.

Juran, J.M. and Gryna, F.M. (1998). Juran's Quality Control Handbook, 4th ed., New York: McGraw-Hill.

Kamarthi, S.V. and Kumara, S.R.T. (1993). Neural Networks in Conceptual Design, In: Neural Networks in Design and Manufacturing, Wang, J. and Takefuji, Y. (eds.), Singapore: World Scientific, pp. 99–120.

Kahn, H., Filer, N., Williams, A. and Whitaker, N. (2001). A Generic Framework for Transforming EXPRESS Information Models, Computer-Aided Design, **33**, pp. 501–510.

Kim, B., Leung, J.M.Y., Park, K.T., Zhang, G. and Lee, S. (2002). Configuring a Manufacturing Firm's Supply Network with Multiple Suppliers, IIE Transactions, **34**(8), pp. 663–677.

Kimberly, W. (1999). Back to the Future, Automotive Engineer, **24**(5), pp. 62–64.

Kimura, F., Hata, T. and Suzuki, H. (1998). Product Quality Evaluation Based on Behaviour Simulation of Used Products, Annals of the CIRP, **47**(1), pp. 119–122.

Kimura, F. and Nielsen, J. (2005). A Design Method for Product Family under Manufacturing Resource Constraints, Annals of the CIRP, **54**(1), pp. 139–142.

King, A.M. and Sivaloganathan, S. (1998). Flexible Design: a Methodology for Strategic Reuse, In: Design Reuse, Engineering Design Conference' 98, Sivaloganathan, S. and Shahin, T.M.M., (eds) (1999). London: Professional Engineering Publishing, pp.337–347

Kirschman, C.F. and Fadel, G.M. (1998). Classifying Functions for Mechanical Design, ASME Journal of Mechanical Design, **120**(3), pp. 475–482.

Kleijnen, J.P.C. (1987). Statistical Tools for Simulation Practitioners, New York: Marcel Dekker.

Kohonen, T. (1989). Self-Organization and Associative Memory, 3rd ed., New York: Springer-Verlag.

Koren, Y. and Kota, S. (1999). Reconfigurable Machine Tools, U.S. Patent No. 5,943,750.

Koren, Y., Heisel, U., Jovane, F., Moriwaki, T., Pritschow, G., Ulsoy, G. and Van Brussel, H. (1999). Reconfigurable Manufacturing Systems, Annals of the CIRP, **48**(2), pp. 527–540.

Kota, S., Sethuraman, K. and Miller, R. (2000). A Metric for Evaluating Design Commonality in Product Families, ASME Journal of Mechanical Design, **122**(4), pp. 403–410.

Kusiak, A. and Tseng, T.L. (2000). Data Mining in Engineering Design: A Case Study, In: Proceedings of the 2000 IEEE International Conference on Robotics & Automation, San Francisco, CA., April, pp. 206–211.

Kusiak, A. and Wang, J. (1995). Dependency Analysis in Constraint Negotiation, IEEE Transactions on Systems, Man, and Cybernetics, **25**(9), pp. 1301–1313.

Lai, K. and Wilson, W.R.D. (1989). FDL - A Language for Function Description and Rationalization in Mechanical Design, ASME Journal of Mechanics, Transmissions, and Automation in Design., **111**, 117–123.

Lander, S.E. (1997). Issues in Multiagent Design Systems, IEEE Expert, **12**(4), pp. 18–26.

Lehnerd, A.P. (1987). Revitalizing the Manufacture and Design of Mature Global Products, In: Technology and Global Industry: Companies and Nations in the World Economy, Guile, B.R. and Brooks, H., (eds.), Washington, DC: National Academy Press, pp. 49–64.

Li, H. and Azarm, S. (2002). An Approach for Product Line Design Selection under Uncertainty and Competition, ASME Journal of Mechanical Design, **124**(3), pp. 385–392.

Li, W.D., Ong, S.K. and Nee, A.Y.C. (2006). Integrated and Collaborative Product Development Environment, Technologies and Implementations, Singapore: World Scientific.

Liang, W.Y. and Huang, C.C. (2002). The Agent-based Collaboration Information System of Product Development, International Journal of Information Management, **22**(3), pp. 211–224.

Liu, J. and Zhong, N. (1999). Intelligent Agent Technology: Systems, Methodologies, and Tools, Proceedings of the 1st Asia-Pacific Conference on Intelligent Agent Technology, Singapore: World Scientific.

MacCallum, K.J. and Duffy A.H.B. (1995). Design Concept Reuse in New Engineering Contexts: DESRU, http://www.cad.strath.ac.uk/research/projects/DESRU.html.

Maher, M.L. and Pu, P. (1997). Issues and Applications of Case-based Reasoning in Design, New Jersey: Lauwrence Erlbaum.

Malmqvist, J. (1994). Computational Synthesis and Simulation of Dynamic Systems, In: ASME Proceedings of the Design Theory and Methodology Conference, New York: ASME, pp. 221–230.

Manfaat, D., Duffy, A.H.B. and Lee, B.S. (1998). SPIDA: Abstracting and Generalizing Layout Design Cases, Artificial Intelligence for Engineering Design, Analysis and Manufacturing, **12**, pp. 141–159.

Mangun, D. and Thurston, D.L. (2002). Incorporating Component Reuse, Remanufacture, and Recycle into Product Portfolio Design, IEEE Transactions on Engineering Management, **49**(4), pp. 479–490.

Martin, M.V. and Ishii, K. (1996). Design for Variety: a Methodology for Understanding the Costs of Product Proliferation, In: Proceedings of the 1996 ASME Design Engineering Technical Conferences, Paper No. 96-DETC/DTM-1610, Irvine, CA, September.

Martin, M.V. and Ishii, K. (1997). Design for Variety: Development of Complexity Indices and Design Charts, In: Proceedings of the 1997 ASME Design Engineering Technical Conferences, Paper No. DETC97/DFM-4359, Sacramento, CA, September.

Martin, M. and Ishii, K. (2002). Design for Variety: Developing Standardized and Modularized Product Platform Architectures, Research in Engineering Design, **13**, pp. 213–235.

Mattson, C.A. and Magleby, S.P. (2001). The Influence of Product Modularity during Concept Selection of Consumer Products, ASME Design Engineering Technical Conferences - Design Theory & Methodology Conference, Paper No. DETC2001/DTM-21712, Pittsburgh, PA., September.

McAdams, D.A., Stone, R.B. and Wood, K.L. (1999). Function Interdependence and Product Similarity Based on Customer Needs, Research in Engineering Design, **11**, pp. 1–19.

McClure, C. (1997). Software Reuse Techniques: Adding Reuse to the System Development Process, Upper Saddle River, NJ: Prentice Hall.

McKay, A., Erens, F. and Bloor, M.S. (1996). Relating Product Definition and Product Variety, Research in Engineering Design, **8**(2), pp. 63–80.

Medina, J.F. and Duffy, M.F. (1998). Standardization vs Globalization: a New Perspective of Brand Strategies, Journal of Product and Brand Management, **7**(3), pp. 223–243.

Messac, A., Martinez, M.P. and Simpson, T.W. (2002). A Penalty Function for Product Family Design Using Physical Programming, ASME Journal of Mechanical Design, **124**(2), pp. 164–172.

Meyer, B. (1994). Reusable Software, the Base Object-Oriented Component Libraries, Hemel Hempstead: Prentice Hall.

Meyer, M.H. and Lehnerd, A.P. (1997). The Power of Product Platforms- Building Value and Cost Leadership, New York: The Free Press.

Michaels, J.V. and Wood, W.P., 1989, Design to Cost, New York: John Wiley & Sons.

Mittal, S. and Frayman, F. (1989). Towards a Generic Model of Configuration Tasks. In: Proceedings of the 11th International Joint Conference on Artificial Intelligence. Detroit, Michigan, August 20-25, pp. 1395–1401.

Monedero, J. (2000). Parametric Design: a Review and Some Experiences, Automation in Construction, **9**, pp. 369–377.

Mori, T. (1990). The New Experimental Design: Taguchi's Approach to Quality Engineering, Dearborn, MI: ASI Press.

Moss, J., Cagan, J. and Kotovsky, K. (2004). Learning from Design Experience in an Agent-based Design System, Research in Engineering Design, **15**, pp. 77–92.

Myers, R.H. and Montgomery, D.C. (2002). Response Surface Methodology: Process and Product Optimization Using Designed Experiments, 2nd ed., New York: John Wiley & Sons.

Nayak, R.U., Chen, W. and Simpson, T.W. (2002). A Variation-Based Method for Product Family Design, Engineering Optimization, **34**(1), pp. 65–81.

Nelson, S.A., II, Parkinson, M.B. and Papalambros, P.Y. (2001). Multicriteria Optimization in Product Platform Design, ASME Journal of Mechanical Design, **123**(2), pp. 199–204.

Ong, S.K., Lin, Q. and Nee, A.Y.C. (2006). Web-Based Configuration Design System for Product Customization, International Journal of Production Research, **44**(2), pp. 351–382.

Ostwald, P.F. and McLaren, T.S. (2004). Cost Analysis and Estimating for Engineering and Management, Upper Saddle River, NJ: Pearson/Prentice Hall.

Otto, K. and Wood, K.L. (2001). Product Design Techniques in Reverse Engineering and New Product Development, NJ: Prentice Hall.

Pahl, G. and Beitz, W. (1996). Engineering Design: a Systematic Approach, 2nd ed., London: Springer.

Park, J. and Simpson, T.W. (2005). Development of a Production Cost Estimation Framework to Support Product Family Design, International Journal of Production Research, **43**(4), pp. 731–772.

Pawlak, Z. (1991). Rough Sets: Theoretical Aspects of Reasoning about Data, Boston, MA.: Kluwer.

Pearce, M., Goel, A.K., Kolonder, J.L., Zimrig, C., Sentisa, L. and Billington R. (1992). Case-Based Design Support - A Case Study in Architectural Design, IEEE Expert, **7**(5), pp. 14–20.

Pena-Mora, F. and Vadhavkar, S. (1996). Augmenting Design Patterns with Design Rationale, Artificial Intelligence for Engineering Design, Analysis and Manufacturing, **11**, pp. 93–108.

Phadke, M.S. (1989). Quality Engineering Using Robust Design, Englewood Cliffs, NJ: Prentice Hall.

Pine, B.J. (1993a). Mass Customization: the New Frontier in Business Competition, Boston: Harvard Business School Press.

Pine, J.B. (1993b). Standard Modules Allow Mass Customization at Bally Engineering Structures, Planning Review, **21**(4), pp. 20–22.

Portter, S., Darlington, M.J., Culley, S.J. and Chawdhry, P.K. (2001). Design Synthesis Knowledge and Inductive Machine Learning, Artificial Intelligence for Engineering Design, Analysis and Manufacturing, **15**, pp. 233–249

Pratt, M.J. and Anderson, B.D. (2001). A Shape Modeling Applications Programming Interface for the STEP Standard, Computer-Aided Design, **33**, pp. 531–543.

Pugh, S. (1991). Total Design, Integrated Methods for Successful Product Engineering, Addison-Wesley Publishing Company.

Pulm, U. and Lindemann, U. (2001). Enhanced Systematics for Functional Product Structuring, Design Research- Theories, Methodologies and Product Modelling, Culley, S. Duffy, A., McMahon, C. and Wallace, K., (eds.), Professional Engineering Publishing, pp. 477–484.

Qian, L. and Gero, J.S. (1996). Function-Behavior-Structure Paths and Their Role in Analogy-Based Design, Artificial Intelligence for Engineering Design, Analysis and Manufacturing, **10**(4), pp. 289–312.

Rada, R. (1995). Software Reuse, NJ.: Ablex Publishing Corporation.

Rai, R. and Allada, V. (2003). Modular Product Family Design: Agent-based Pareto-optimization and Quality Loss Function-Based Post-optimal Analysis, International Journal of Production Research, **41**(17), pp. 4075–4098.

Regli, V.V. and Cicirello, V.A. (2000). Managing Digital Libraries for Computer-Aided Design, Computer-Aided Design, **32**(2), pp.119–132

Reich, Y. (1993). The Development of BRIDGER: A Methodological Study of Research on the Use of Machine Learning in Design, Artificial Intelligence in Engineering, **8**(3), pp.217–231.

Reich, Y., Konda, S.L., Levy, S.N., Monarch, I.A. and Subrahmanian, E. (1993). New Roles for Machine Learning in Design, Artificial Intelligence in Engineering, **8**, pp. 165–181.

Rezayat, M. (2000). Knowledge-Based Product Development Using XML and KCs, Computer-Aided Design, **32**(5), pp. 299–309.

Rolstadas, A. (1991). Editorial: Group Technology and Design of Production Systems, Production Planning and Control, **2**(4), pp. 297.

Romanowski, C.J., Nagi, R. (2002). A Data Mining and Graph Theoretic Approach to Building Generic Bills of Materials, In: 11th International Engineering Research Conference, Orlando, FL., May.

Roy, U., Pramanik, N., Sudarsan, R., Sriram, R.D. and Lyons, K.W. (2001). Function-to-Form Mapping: Model, Representation and Applications in Design Synthesis, Computer-Aided Design, **33**, pp. 699–719.

Rumelhart, D.E., Widrow, B. and Lehr, M.A. (1994). The Basic Ideas in Neural Networks, Communications of the ACM, **37**(3), pp87–92.

Rychener, M.D. (1989). Expert Systems for Engineering Design, New York: Academic Press Inc.

Sabbagh, K. (1996). Twenty-first Century Jet: the Making and Marketing of the Boeing 777, New York: Scribner.

Sand, J.C., Gu, P. and Watson, G. (2002). HOME: House of Modular Enhancement - A Tool for Modular Product Redesign, Concurrent Engineering: Research and Applications, 10(2), pp. 153–164.

Sarker, B.R. and Islam, K.M.S. (1999). Relative Performance of Similarity and Dissimilarity Measures, Computers and Industrial Engineering, 37, pp. 769–807.

Savransky, S.D. (2000). Engineering of Creativity: Introduction to TRIZ Methodology of Inventive Problem Solving, Boca Raton, Fla.: CRC Press.

Schaffer, J. (1985). Multiple Objective Optimization with Vector Evaluated Genetic Algorithms, In: Genetic Algorithms and Their Applications: Proceedings of 1st International Conference on Genetic Algorithms, Grefenstette, J.J., (ed.), Princeton, NJ: Lawrence Erlbaum, pp. 93–100.

Schniederjans, M.J. (1995). Goal Programming: Methodology and Applications, Boston: Kluwer Academic Publishers.

Scott, J. and Sen, P. (1998). Focusing Design Reuse Initiatives Using the Design Structure Matrix System, In: Design Reuse, Engineering Design Conference' 98, Sivaloganathan, S. and Shahin, T.M.M., (eds), 1999, London: Professional Engineering Publishing, pp. 351–359.

Shahin, T.M.M. (1997). Automation of Feature-based Modelling & Finite Element Analysis for Optimal Design, PhD thesis, Brunel University, UK.

Shen, W., Norrie, D.H. and Barthès, J.P. (2001). Multi-agent Systems for Concurrent Intelligent Design and Manufacturing, New York: Taylor & Francis.

Shipman, F. and McCall, R. (1996). Integrating Different Perspectives on Design Rationale: Supporting the Emergence of Design Rationale from Design Communication, Artificial Intelligence for Engineering Design, Analysis, and Manufacturing, 11, pp. 141–154.

Shuford, R.H. (1995). Activity-Based Costing and Traditional Cost Allocation Structures, In: Cost Estimator's Reference Manual, 2nd ed., Stewart, R., Wyskida, R. and Johannes, J., (eds.), New York: John Wiley & Sons (1995). pp. 41–94.

Siddall, J.N. (1983). Probabilistic Engineering Design: Principles and Applications, New York: M. Dekker.

Siddique, Z. (2000). Common Platform Development: Design for Product Variety, Ph.D. Dissertation, Georgia Institute of Technology, Atlanta, GA.

Siddique, Z. and Rosen D.W. (2001). On Combinatorial Design Spaces for the Configuration Design of Product Families, Artificial Intelligence for Engineering Design, Analysis and Manufacturing, 15(2), pp. 91–108.

Sim, S.K. and Duffy, A.H.B. (1998). A Foundation for Machine Learning in Design, Artificial intelligence for Engineering Design, Analysis and Manufacturing, 12, pp. 193–209.

Simpson, T.W. (1998). A Concept Exploration Method for Product Family Design, Ph.D. Dissertation, Georgia Institute of Technology, Atlanta, GA.

Simpson, T.W. (2004). Product Platform Design and Customization: Status and Promise, Artificial Intelligence for Engineering Design, Analysis and Manufacturing, **18**, pp. 3–20.

Simpson, T.W., Chen, W, Allen, J.K. and Mistree, F. (1996). Conceptual Design of a Family of Products through the Use of the Robust Concept Exploration Method, In: Proceedings of the 6th AIAA/USAF/NASA/ISSMO Symposium on Multidisciplinary Analysis and Optimization, Published by AIAA, Inc., **2**(2), pp. 1535–1545.

Simpson, T.W., Chen, W., Allen, J.K. and Mistree, F. (1997a). Designing Ranged Sets of Top-Level Design Specifications for a Family of Aircraft: An Application of Design Capability Indices, In: Proceedings of the SAE World Aviation Congress and Exposition, Paper No. AIAA-97-5513, Anaheim, CA, October.

Simpson, T. W., Peplinski, J., Koch, P. N. and Allen, J. K. (1997b). On the Use of Statistics in Design and the Implications for Deterministic Computer Experiments, In: Proceedings of 1997 ASME Design Engineering Technical Conferences, Paper No. DETC97/DTM-3881, Sacramento, CA, September.

Simpson, T. W., Maier, J.R.A. and Mistree, F. (2001). Product Platform Design: Method and Application, Research in Engineering Design, **13**(1), pp. 2–22.

Sivaloganathan, S. and Shahin, T.M.M. (1999). Design Reuse: An Overview, In: Proceedings of the Institute of Mechanical Engineers, Part B: Journal of Engineering Manufacture, **213**(7). pp. 641–654.

Smith, J.S. (2002). A Multiple Viewpoint Modular Design Methodology, Ph.D. Dissertation, Department of Design, Manufacturing and Engineering Management, University of Strathclyde, UK.

Srinivas, N. and Deb, K. (1995). Multi-Objective Function Optimization Using Non-dominated Sorting Genetic Algorithms, Evolutionary Computation, **2**, pp. 221–248.

Stahovich, T.F. (2000). LearnIT: An Instance-based Approach to Learning and Reusing Design Strategies, ASME, Journal of Mechanical Design, **122**, pp. 249–256.

Stake, R.B. and Blackenfelt, M. (1998). Modularity in Use – Experiences from Five Companies, 4th WDK Workshop on Product Structuring, Delft, Netherlands, October 22–23.

Steward, D.V. (1981). Systems Analysis and Management, Structure, Strategy, and Design, New York: Petrocelli Books Inc.

Stone, R.B., Wood, K.L. and Crawford, R.H. (2000a). Using Quantitative Functional Models to Develop Product Architectures, Design Studies, **21**(3), pp. 239–260.

Stone, R.B., Wood, K.L. and Crawford, R.H. (2000b). A Heuristic Method for Identifying Modules for Product Architectures, Design Studies, **21**(1), pp. 5–31.

Stone, R. and Wood, K.L. (2000). Development of a Functional Basis for Design, Journal of Mechanical Design, **122**, pp. 359–370.

Stork, A., Brunetti, G. and Vieira, A.S. (1996). Intuitive Semantically Constrained Interaction in Feature-based Parametric Design, In: Proceedings of CAD'96, Distributed and Intelligent CAD Systems, pp. 77–91.

Sundgren, N. (1999). Introducing Interface Management in New Product Family Development, Journal of Product Innovation Management, 16(1), pp. 40–51.

Su, D. (1998). Application of Expert System and Artificial Neural Networks for Design Knowledge Retrieval, In: Design Reuse, Engineering Design Conference' 98, Sivaloganathan, S. and Shahin, T.M.M., (eds), 1999, London: Professional Engineering Publishing, pp.371–378.

Suh, N.P. (2001). Axiomatic Design- Advances and Applications, New York: Oxford University Press.

Suh, N.P. (2005). Complexity: Theory and Applications, New York: Oxford University Press.

Szykman, S., Sriram, R.D. Bochenek, C. and Racz, J. (1998). The NIST Design Repository Project, Advances in Soft Computing – Engineering Design and Manufacturing, Roy, R., Furuhashi, T. and Chawdhry, P.K., (eds.), London: Springer-Verlag, pp. 5–19.

Szykman, S., Racz, J.W. and Sriram, R.D. (1999). The Representation of Function in Computer-Based Design, In: Proceedings of the 1999 ASME Design Engineering Technical Conferences, Paper No. DETC99/DFM-8742, Las Vegas, Nevada, September.

Szykman, S., Fenves, S.J., Keirouz, W. and Shooter, S.B. (2001). A Foundation for Interoperability in Next Generation Product Development Systems, Computer-Aided Design, 33, pp. 545–559.

Taguchi, G. (1986). Introduction to Quality Engineering: Designing Quality into Products and Processes, Tokyo: Assian Productivity Organization.

Treleven, M. and Wacker, J.G. (1987). The Sources, Measurements, and Managerial Implications of Process Commonality, Journal of Operations Management, 7, pp. 11–25.

Ullman, D.G. (1997). The Mechanical Design Process, New York: McGraw-Hill.

Ulrich, K. (1995). The Role of Product Architecture in the Manufacturing Firm, Research Policy, 24, pp. 419–440.

Ulrich, K. and Seering, W. (1987). Conceptual Design: Synthesis of Systems of Components, In: Proceedings of the 1987 ASME Winter Annual Meeting Symposium on Integrated and Intelligent Manufacturing, Boston, 1987, pp. 57–66.

Ulrich, K. and Seering, W. (1989). Synthesis of Schematic Descriptions in Mechanical Design, Research In Engineering Design, 1(1), pp. 3–18.

Ulrich, K. and Eppinger, S. (2004). Product Design and Development, 3rd ed., Boston: McGraw-Hill/Irwin.

Umeda, Y., Takeda, H., Tomiyama, T. and Yoshikawa, H. (1990). Function, Behavior and Structure, In: Proceedings of the AIENG'90 Applications of AI in Engineering, Boston, July, pp. 177–193.

Uzumeri, M. and Sanderson, S. (1995). A Framework for Model and Product Family Competition, Research Policy, **24**: pp. 583–607.

Van Wie, M.J., Greer, J.L., Campbell, M.I., Stone, R.B. and Wood, K.L. (2001). Interfaces and Product Architecture, In: Proceedings of ASME 2001 Design Engineering Technical Conferences and Computers and Information in Engineering Conferences, Paper No. DETC01/DTM-21689, Pittsburgh, Penn., September.

Vjekoslav, D. and Montgomery, J. (2003). Mechatronics by Bond Graphs: an Object-Oriented Approach to Modelling and Simulation, Berlin, New York: Springer.

Wacker, J.G. and Treleven, M. (1986). Component Part Standardization: An Analysis of Commonality Sources and Indices, Journal of Operations Management, **6**, pp. 219–244.

Wang, Z.G., Wong, Y.S. and Rahman, M. (2005). Development of a Parallel Optimization Method Based on Genetic Simulated Annealing Algorithm, Parallel Computing, **37**(8,9), pp. 839–857.

Warfield, J.N. (1973). Binary Matrices in System Modeling, IEEE Transactions on Systems, Man, and Cybernetics, **3**(5), pp. 441–449.

Wasmund, M. (1993). Implementing Critical Success Factors in Software Reuse, IBM Systems Journal, **32**(4), pp. 595–611.

Watson, I. (1999). Case-Based Reasoning is a Methodology not a Technology, Knowledge-Based Systems, **12**, pp. 303–308.

Watson, I. and Perera, S. (1997). Case-based Design: A Review and Analysis of Building Design Applications, Artificial Intelligence for Engineering Design, Analysis and Manufacturing, **11**(1), pp. 59–87.

Whitney, D.E. (1993). Nippondenso Co. Ltd: A Case Study of Strategic Product Design, Research in Engineering Design, **5**(1), pp. 1–20.

Wielinga, B. and Schreiber, G. (1997). Configuration-Design Problem Solving, IEEE Intelligent Systems, **12**(2), pp. 49–56.

Wilhelm, B. (1997). Platform and Modular Concepts at Volkswagen - Their Effect on the Assembly Process, In: Transforming Automobile Assembly: Experience in Automation and Work Organization, Shimokawa, K., Jürgens, U. and Fujimoto, T. (eds.), New York: Springer-Verlag, pp. 146–156.

Wortmann, J.C., Muntslag, D.R. and Timmermans, P.J.M. (1997). Customer Driven Manufacturing, London: Chapman & Hall.

Yen, J., Liao, J.C. and Lee, B. (1998). A Hybrid Approach to Modelling Metabolic Systems Using Genetic Algorithm and Simplex Method, IEEE Transactions on Systems, Man, and Cybernetics, Part B, **28**(2), pp. 173–191.

Yu, T., Yassine, A. and Goldberg, D. (2003). A Genetic Algorithm for Developing Modular Product Architectures, In: Proceedings of ASME 2003 Design Engineering Technical Conferences and Computers and Information in Engineering Conferences, Paper No. DETC2003/DTM-48657, Chicago, Illinois, September.

Yu, J.S., Gonzalez-Zugasti, J.P. and Otto, K.N. (1999). Product Architecture Definition Based upon Customer Demands, ASME Journal of Mechanical Design, **121**(3), pp. 329–335.

Yu, B. and MacCallum, K. (1995). Modelling of Product Configuration Design and Management by Using Product Structure Knowledge, In: Knowledge Intensive CAD, Volume 1, Tomiyama, T., Mantyla, M. and Finger, S. (eds.), London: Chapman & Hall, pp. 115–124.

Zalzala, A.M.S. and Fleming, P.J. (1997). Genetic Algorithms in Engineering Systems, London: Institution of Electrical Engineers.

Zamirowski, E.J. and Otto, K.N. (1999). Identifying Product Family Architecture Modularity Using Function and Variety Heuristics, In: Proceedings of the ASME Design Engineering Technical Conferences, Paper No. DETC99/ DTM-8760, Las Vegas, Nevada.

Zhang, F. and Xue, D. (2001). Optimal Concurrent Design Based upon Distributed Product Development Life-cycle Modeling, Robotics and Computer-Integrated Manufacturing, **17,** pp. 469–486.

Zhong, N., Liu, L., Ohsuga, S. and Bradshaw, J. (2001). Intelligent Agent Technology: Systems, Methodologies, and Tools, Proceedings of the 2nd Asia-Pacific Conference on Intelligent Agent Technology, Singapore: World Scientific.

Zhou, Z.D., Wang, H.H., Chen, Y.P., Ai, W., Ong, S.K., Fuh, J.Y.H. and Nee, A.Y.C. (2003). A Multi-Agent-Based Agile Scheduling Model for a Virtual Manufacturing Environment, International Journal of Advanced Manufacturing Technology, **21**(12), pp.980–984.

Index

Activity-based costing, 163, 167
  ABC, 163, 164, 165, 166, 170
Adaptable design, 34
A-Design, 21, 30, 51, 130, 131, 211
Agent-based, 15, 50, 51, 122, 130, 131, 211
Application Programming Interface
  API, 249, 250, 251, 252
Artificial Intelligence, 1
  AI, 1, 10, 14, 21, 36, 37, 40, 51, 52, 56, 217
Artificial Neural Network, 46
  ANN, 46, 47, 48, 111
Atomic function, 69, 70, 72, 73, 78, 80, 82, 95, 97, 101, 102, 103, 105
Automated design synthesis, 15, 21, 30, 51, 118, 122, 158, 201, 212, 222, 223, 240
Axiomatic design, 6, 46, 111, 175, 184, 200, 206
  independence axiom, 184
  information axiom, 184

Back-Propagation, 47, 92
  BP, 92
Behavior, 13, 38, 56, 57, 62, 68, 124, 176, 207
Bill-of-materials, 22
  BOM, 22, 68, 168
Black & Decker, 8, 33
Boeing, 9

Bond graph, 124
Branch-and-bound, 15, 122, 211

Case-based reasoning, 27, 28
  CBR, 12, 28, 29, 31, 38, 237, 242
CASECAD, 29
Catalog-based design, 29, 30, 31
Chromosomes, 136, 245
Client-server, 218
Commonality, 8, 9, 82, 84, 85, 87, 168, 169, 175, 214, 230, 237
  index, 168, 169, 175
Component capability index, 115, 175, 206, 216
Component catalog, 30, 31, 112, 114, 115, 116, 131, 135, 142, 150, 188, 190, 216
Compositional modeling language, 13, 60
  CML, 13, 60
Computer-Aided Design, 1
  CAD, 1, 25, 54, 59, 61, 65, 67, 70, 79, 82, 249, 250, 253
Computer-Aided Engineering, 1
  CAE, 1
Computer-Aided Manufacturing
  CAM, 1, 54
Computer-Aided Process Planning, 1
  CAPP, 1
Conceptual design, 6, 7, 8, 13, 21, 29, 37, 160, 171, 182, 237

Concurrent Engineering, 51, 130
Configuration design, 15, 50, 113, 120, 122, 123, 125, 126, 127, 128, 129, 130, 131, 136, 143, 158, 178, 206, 207, 212, 231, 240
Constraint satisfaction problem, 123
    CSP, 123
Correlation matrix, 79, 191, 194, 215, 223, 232
Cost
    Cost center, 164, 165
    Cost driver, 163, 164, 165, 166, 167
    Cost estimating relationships, 161
    Cost estimation, 37, 153, 154, 155, 156, 157, 160, 162, 163, 167, 168, 169, 170, 171, 175, 207
    Cost model, 153, 155, 158, 159, 161, 166, 167, 169, 170, 171, 172, 173, 175, 215, 216, 234, 235, 237, 239
    Cost road-map, 115, 174, 215, 225, 234, 239
    Cost structure, 155, 159, 175
Crossover, 49, 119, 127, 135, 245

Data mining, 14, 22, 42, 43, 44
Database management systems, 66
    DBMS, 66, 67, 218
Decision-making, 5, 6, 21, 30, 39, 48, 51, 53, 56, 115, 166, 205, 239
Decomposition, 17, 37, 41, 69, 78, 87, 90, 101, 111, 171, 214, 219, 221, 252
Degree of commonality index, 168
    DCI, 168
Design constraint, 127, 132, 135, 141, 147, 150, 201, 234, 238, 239
Design for variety, 32
    DFV, 32, 88
Design freedom, 6, 7, 54, 86
Design of experiments, 177, 206
    DOE, 177, 182, 183, 206, 207
Design parameters, 81, 112, 167, 175, 176, 215

DPs, 81, 84, 111, 112, 115, 114, 131, 135, 159, 160, 161, 175, 176, 177, 178, 182, 183, 184, 185, 199, 206, 215, 254
Design range, 141, 178, 185, 186, 187, 195, 196, 197, 203
design reuse, 1, 2, 3, 4, 5, 6, 10, 11, 12, 13, 14, 16, 18, 19, 20, 21, 25, 26, 27, 28, 27, 28, 29, 31, 37, 40, 42, 45, 52, 53, 81, 113, 114, 143, 146, 148, 150, 157, 158, 153, 171, 175, 176, 182, 199, 200, 205, 206, 205, 206, 212, 215, 217, 223, 238, 239, 237, 238, 239, 240, 241, 242, 245, 249, 251, 252, 253
Design reuse, 2, 10, 11, 16, 20, 21, 27, 51, 81, 205, 212, 213
Design reuse process model, 11, 12, 27
Design structure matrix, 11, 44, 88
    DSM, 11, 32, 44, 45, 46, 88, 89, 110, 111
Design synthesis, 14, 15, 31, 42, 50, 51, 86, 113, 115, 113, 120, 122, 123, 124, 126, 131, 133, 134, 136, 137, 141, 142, 143, 145, 150, 151, 157, 158, 153, 187, 196, 197, 199, 201, 202, 205, 206, 207, 211, 213, 216, 217, 221, 222, 231, 236, 238, 240
Design to cost, 167
    DTC, 167
Design-by-analogy, 183, 199
Design-by-reuse, 12, 81, 113, 239, 238
Design-for-reuse, 12, 238
Detailed design, 6, 28, 45, 54, 167, 239, 237, 239, 241, 248, 249, 250, 252
Direct cost, 156

Embodiment design, 6, 237, 239, 241, 252
EXPRESS, 60, 62
eXtensible Markup Language, 13
    XML, 13, 60, 208

Fan filter unit, 27, 145

FFU, 27, 28, 145, 146, 147, 150, 158, 200, 202
Feature, 43, 55, 60, 70, 92, 93, 95, 99, 105, 107, 110, 150, 178, 187, 200, 204, 231, 238, 243, 249, 250, 251, 252
Feature map, 92, 93, 95, 99, 105, 107, 110, 150
Feature-based model, 250
Finite element analysis, 176
  FEA, 176
Flexible manufacturing systems, 23
  FMS, 23, 24
Flow, 12, 17, 31, 55, 56, 64, 65, 69, 71, 72, 73, 74, 75, 76, 77, 84, 90, 96, 124, 154, 155, 241, 244
  Energy flow, 96, 244
  Flow taxonomy, 65, 72, 74, 76
  Material flow, 96, 244
  Signal flow, 96
Foreign key, 218
  FK, 218
Function
  FAST, 69
  FPA, 81, 92, 101, 109, 111, 112, 115, 118, 150, 214, 223
  Function analysis, 92, 100, 107, 110, 111, 214, 239
  Function analysis system technique, 69
  Function clustering, 95, 96, 97
  Function costing, 160, 167, 170
  Function structure, 13, 55, 69, 72, 73, 78, 79, 90, 95, 101, 111, 214, 219, 223, 241
  Function taxonomy, 72, 75, 241, 242
  Function-based, 13, 32, 51, 55, 56, 63, 83, 81, 124, 153, 170, 212, 214
  Function-based product architecture, 81, 153, 214
  Function-form-behavior, 57
Functional requirement, 43, 81, 184, 250, 252

General design space, 113, 213, 214, 215
Generic product modeling, 58
  GPM, 58
Genetic Algorithm, 15, 48, 49, 50, 113, 118, 131, 211, 245
  GA, 15, 49, 89, 91, 113, 118, 119, 122, 126, 127, 129, 131, 135, 139, 158, 207, 208, 210, 211, 212, 237, 243, 244, 245, 252
Goal programming, 116
  GP, 116, 117, 158
Group Technology, 23
  GT, 23, 24

HEPA, 149, 157, 202
HOMER, 29
Host product, 141, 191, 192, 200, 226
House of modular enhancement
  HOME, 88
House of quality, 87, 215
  HOQ, 87, 88, 110, 183, 184, 215
HyperText Markup Language, 249
  HTML, 252
HyperText Transfer Protocol
  HTTP, 249, 250, 252

IDEF0, 61
Independence axiom, 184
Indirect cost, 156, 162, 163, 164, 166, 167
Information
  ICA, 175, 187, 188, 189, 195, 199, 200, 205, 206, 213, 222, 238, 239
  Information axiom, 184
  Information content, 175, 178, 184, 185, 186, 187, 195, 196, 197, 198, 200, 201, 202, 203, 204, 206, 213, 216, 226, 227, 228, 230, 231, 234, 236, 238, 239, 240
  Information Content Assessment, 175, 182, 187, 213
  Information model, 12, 31, 42, 51, 53, 55, 60, 61, 62, 63, 66, 69, 83, 205, 206, 207, 212, 214, 217

Information modeling, 12, 42, 60, 61, 62, 63, 66, 205, 206, 207, 212, 217
Initial graphics exchange specification, 59
IGES, 59
Internet, 60, 65, 219, 239, 249, 250, 252
ISO 10303, 61, 62

Key characteristic, 57, 154, 213, 231, 241
KC, 69, 70, 79, 117, 154, 188, 189, 191, 192, 196, 197, 200, 203, 204, 205, 213, 215, 218, 226, 234
Key element vector, 72, 73, 95
KEV, 73, 78, 84, 96, 101, 109
Knowledge extraction operator, 111, 213, 214, 216, 221
Knowledge-based, 36, 58, 65
Kohonen model, 93, 94, 95

Learning curve, 62, 157, 158, 159, 225
Learning effect, 157, 170, 174, 225
Life-cycle, 18, 182, 238
Linear programming, 50, 125, 126, 212
LP, 125, 126, 127

Machine learning, 14, 29, 40, 41, 42, 44, 212
Magnitude-based costing, 162
MBC, 162, 170
Management science, 27, 168
Manufacturability, 32
Manufacturing feature, 48
Mapping route, 112, 115, 116, 118, 187, 215
Mass customization, 82, 163, 168, 178
Meta-model, 176, 177, 193, 199, 206, 207
Meta-modeling, 177, 207
Modular design, 8, 9, 20, 31, 32, 33, 34, 35, 86, 110, 128, 173, 223
Modular function deployment, 32, 88
MFD, 32, 88, 89

Modularity, 8, 32, 33, 87, 88, 91, 169, 212
Module-based, 83, 86, 87, 117, 130, 206, 209, 211
Multi-Agent Systems, 50, 130
MAS, 50, 51
Multi-level programming, 118, 158
Multi-objective GA, 119
MOGA, 120
multi-objective struggle genetic algorithm, 113, 213, 216
Multi-objective struggle genetic algorithm
MOSGA, 113, 119, 133, 134, 136, 142, 143, 158, 201, 213, 216, 222, 226, 236
Multiple-objective optimization, 216
Multiple-objective optimization problem, 216
MOOP, 113, 115, 116, 117, 118, 132, 144, 145, 216, 236
Mutation, 49, 119, 127, 135, 245
MySQL, 66

Neural networks toolbox, 100
Neurons, 46, 93, 94, 95, 98
NIST design repository project, 30, 56, 211
NODES, 37, 41
Non-dominated sorting GA, 119
NSGA, 119
Non-linear programming, 125
NLP, 125, 126, 127, 129, 210, 212

Object-oriented, 17, 62, 212
Oracle, 66
Orthogonal arrays, 182

Parameter space, 114, 115, 137, 145
Parametric costing, 161, 167
Pareto set, 116, 117, 144, 157
Pareto-front, 115, 116, 118, 119, 136, 141, 144, 145, 150

Index

Pareto-optimal, 114, 115, 119, 130, 135, 136, 139, 143, 144, 145, 151, 156, 157, 158, 216, 226
Pareto-optimality, 119, 145
P-Diagram, 179, 180
Performance evaluation, 175, 178, 182, 187, 199, 201, 206, 212, 215, 216, 239, 254
PERSPECT, 38, 41
Post-optimal solution selection, 143, 144, 145
Probability density function, 185
  pdf, 185, 192, 196
Probability mass function, 192
  pmf, 192
Product
  PFA, 59
  PFCT, 58
  PFDR, 69, 206, 212, 213, 214, 217, 218, 219, 221, 223, 228, 231, 234, 235, 236, 237, 238, 239
  PLC, 1, 2, 61, 179
  PLM, 68
  PPCEM, 85
  Product architecture, 14, 30, 31, 32, 56, 58, 82, 83, 86, 87, 88, 90, 91, 92, 101, 110, 111, 118, 168, 171, 206, 211, 212, 214, 232
  Product data management, 1, 66, 67, 251
  PDM, 1, 22, 66, 67, 68
  Product family, 7, 8, 9, 10, 28, 31, 58, 69, 83, 84, 85, 86, 88, 92, 99, 106, 111, 112, 113, 123, 124, 127, 128, 130, 131, 143, 153, 168, 169, 170, 171, 172, 173, 175, 195, 200, 205, 207, 211, 212, 214, 215, 216, 217, 221, 222, 223, 225, 226, 227, 228, 230, 231, 234, 235, 236, 237, 239, 240, 241, 242, 243, 244, 245, 246, 249, 251, 252
  Product family architecture, 59
  Product family classification tree, 58

Product family design reuse, 28, 69, 205, 206, 217
Product information model, 12, 31, 53, 56, 65, 68, 69, 79, 83, 92, 95, 107, 116, 187, 212, 213, 220, 241
Product life-cycle, 1, 3, 7, 251
Product platform, 9, 81, 82, 83, 84, 85, 86, 107, 110, 111, 114, 117, 118, 122, 132, 135, 142, 150, 155, 214, 216, 217
Product platform concept exploration method, 85
Productive hour costing, 162, 167
Programmed Attribute Graph Grammar, 59, 124, 212
  PAGG, 59, 124, 209, 212

Quality engineering, 16, 178, 180, 182, 183, 206
Quality function deployment, 16, 87, 215
  QFD, 16, 32, 87, 88, 110, 116, 178, 182, 183, 210, 215
Quality loss function, 180, 183, 184

Radial-basis function, 92
  RBF, 92
Reconfigurable manufacturing systems, 24
Regression analysis, 158, 161, 177
Research and development, 154
Response surface method, 177, 206
  RSM, 177, 182, 183, 206, 207, 208, 211
Robust concept exploration method, 84
  RCEM, 84
Robust design, 6, 87, 175, 178, 179, 180, 182, 206
RODEO, 29
Rule-based, 48

Self organization, 92
Self-organizing map, 92, 94

SOM, 92, 93, 95, 96, 97, 100, 101, 107, 109, 110, 111, 118, 150, 153, 214
Semantics, 13, 52, 60
Signal-to-noise, 182
  S/N, 182, 183, 184
Simulated Annealing, 15, 122, 128, 211
  SA, 15, 91, 122, 126, 128, 129, 131, 209, 210, 211, 212
Single-objective optimization problem, 115
  SOOP, 115, 144, 145
SolidWorks, 79
Sony, 8, 34
SPIDA, 29
Standard for the Exchange of Product model data
  STEP, 13, 60, 61, 62
Stereolithography, 59
  SLS, 59
Swatch, 9, 34
System range, 142, 178, 185, 186, 187, 193, 194, 196, 197, 198, 203, 206

Taguchi method, 16, 178, 180, 182, 184
Taxonomy, 13, 63, 64, 74, 75, 78, 207, 219, 241
  Flow taxonomy, 65, 72, 74, 76

Function taxonomy, 72, 75, 241, 242
Total constant commonality index, 168
  TCCI, 168
TRIZ, 6, 39, 40
TV receiver circuits, 209, 223, 231

Unified Modeling Language, 13, 62
  UML, 13, 60, 62
Unigraphics, 249
  UG, 249, 250, 251, 252
Unsupervised learning, 92, 107, 110, 118

Variant-based platform design methodology, 85
  VBPDM, 85
Vector evaluating genetic algorithm, 119
  VEGA, 119
Virtual Reality Mark-up Language
  VRML, 59, 250, 252

Web-based, 53, 65, 66, 68, 218, 219, 239, 237, 239, 250, 251, 252, 253
Weighted sum method, 116, 144
World-Wide-Web, 65

Xerox, 9